The Institute of Biology's
Studies in Biology no. 25

Animal Photoperiodism

by *Brian Lofts* Ph.D., D.Sc.
Professor of Zoology,
University of Hong Kong

Edward Arnold (Publishers) Ltd

First published 1970

Boards edition SBN : 7131 2293 5
Paper edition SBN : 7131 2294 3

Printed in Great Britain by
William Clowes and Sons Ltd, London and Beccles

General Preface to the Series

It is no longer possible for one textbook to cover the whole field of Biology and to remain sufficiently up to date. At the same time students at school, and indeed those in their first year at universities, must be contemporary in their biological outlook and know where the most important developments are taking place.

The Biological Education Committee, set up jointly by the Royal Society and the Institute of Biology, is sponsoring, therefore, the production of a series of booklets dealing with limited biological topics in which recent progress has been most rapid and important.

A feature of the series is that the booklets indicate as clearly as possible the methods that have been employed in elucidating the problems with which they deal. Wherever appropriate there are suggestions for practical work for the student. To ensure that each booklet is kept up to date, comments and questions about the contents may be sent to the author or the Institute.

1969

INSTITUTE OF BIOLOGY
41 Queen's Gate
London, S.W.7

Preface

Biological rhythms have been known to man since earliest times, and modern culture is replete with practices and customs that reflect a knowledge of the cyclic events which occur daily or seasonally in the lives of animals. The capacity to time annual events such as migration, reproduction, or hibernation must have been of supreme importance at all stages of organic evolution, and it is, therefore, not surprising that animals have evolved mechanisms by which these events can be synchronized to changes in their environment. Many species make use of the daily light cycle as a source of predictive information, and the study of the biological effects of daylength constitutes the subject matter of this book.

As a field of scientific enquiry photoperiodism has expanded enormously during the past thirty years, and has received an added impetus through the increased interest in biological rhythms stimulated by aeronautical and space research. There is now an extensive literature on photoperiodism, for it is a widespread phenomenon found in both vertebrate and invertebrate species. This book is not a review of all these scientific enquiries, rather, it is an introduction to the subject, providing the student with a fundamental understanding of the significance of the phenomenon, and the role it plays in the lives of many species of animals.

Hong Kong, 1970   B.L.

Contents

Animal Cycles and Environmental Synchronizers   1

1.1   Animal cycles

Man has been aware since the very earliest times that animals and plants display annually and diurnally recurring cycles. Primitive man in his constant hunt for food must have noted that many animals produced their young at definite times of the year, and wondered why certain fishes, birds and mammals were only present for part of the year, then suddenly disappeared from their normal localities only to mysteriously reappear many months later. Similarly too, he must have noted the seasonal availability of many fruits and berries. Recognizable drawings of migratory animals have been identified from the crude drawings that decorated the cave dwellings of Neolithic man, and there are numerous biblical references to bird migration. There is no doubt, for example, that the saving of the Israelites from starvation during their wanderings in the wilderness by '. . . a wind from the Lord, that brought quails from the sea . . .', recorded in the Book of Numbers XI.31, refers to a flight of *Coturnix* blown off course during their annual migration. Aristotle and other Grecian scholars wrote of the breeding rhythms displayed by many animals and recorded the seasonal changes in the morphological appearance of the reproductive organs. For centuries the natives of many Pacific islands have held an annual Palolo feast which is dependent upon this marine worm leaving its sandy burrow and appearing in swarms of countless millions in the seas for only a brief period of a few hours in October or November, in the third quarter of the moon. The ancient literature is replete with records of such events.

Equally obvious too, must have been the daily patterns of behavioural activities such as animal feeding rhythms. Our primitive hunter must have soon learned that some animals only emerged and fed at night while others tended to be more active during daylight hours. His very success as a hunter would have been dependent on such knowledge. An empirical knowledge of animal cycles is, therefore, ancient and almost ubiquitous in human culture.

Rhythmicity is a basic characteristic of organic life. Animal evolution has taken place in an environment subject to regular and cyclic fluctuations both of short duration, such as the daily cycles of daylight and darkness, high and low tides, and of much longer periodicities as exemplified by the alternation of spring and neap tides and the gradual changes of the seasons. It is perhaps not surprising to find, therefore, that the survival of a species of plant or animal has required the adaptation of many of their physiological and behavioural processes to these cyclic phenomena. The capacity to bring the physiological mechanisms regulating, for example, reproduction, into relationship with the most beneficial of these changes must have been of prime importance at all stages of evolution, since the maintenance of a species depends on the survival of progeny to an age when they are capable of reproduction. Thus, many of the important events in animals and plants have become orientated to the fluctuations in their respective habitats. Hippocrates recognized the importance of these oscillations, and wrote 'whoever wishes to investigate medicine properly, should proceed thus; in the first place consider the seasons of the year, and what effect each of them produces, for they are not all alike, but differ much from themselves in regard to their changes'.

1.2   Proximate and ultimate factors

Natural selection will strongly favour the gene complexes of those individuals producing offspring at the most propitious season; progeny reared at less favourable times will suffer a high and wasteful mortality. Such a differential survival rate will rapidly define the characteristic breeding season of any particular species, and has, in many of the higher vertebrates at least, resulted in the establishment of annual cycles. BAKER has distinguished two groups of factors which influence these events. The *ultimate factors* are those that exert a selective pressure ensuring that populations breed at the optimal season. In most animals the most important ultimate factor is the availability of suitable food supplies which will determine the number of young that can be reared to a reproductive age. Egg-laying, or the birth of young, is thus usually confined to a period when food for rearing the young is most plentiful. However, it must be appreciated that a multitude of other selective influences impose numerous compromises and adaptations on this basic principle. In the wood-pigeon *Columba palumbus*, for instance, eggs are first laid in the Cambridgeshire area around late April, and continue to be laid right up to late September. Thus, feeding conditions suitable for actual egg-laying continue to prevail until comparatively late in the year. However, towards the latter end of the season, when the days are shorter, the birds may have insufficient time after feeding to indulge in a proper incubation routine, so that the eggs remain unguarded and are exposed for longer periods to predation. As a consequence, fewer eggs give rise to surviving progeny.

In order to 'anticipate' the approach of a suitable season for breeding, animals have evolved response mechanisms to various appropriate environmental stimuli. These are the *proximate factors* in response to which the reproductive organs undergo their physiological development from a seasonally quiescent state to a functional breeding condition. To a large extent the period between conception and the date of hatching or birth is fixed for each species, thus the proximate factors may have to act many months prior to the production of young for the latter to be produced at the opportune time of year. Their ability to serve as effective timing mechanisms is, of course, a product of natural selection, and the particular environmental synchronizer to which an animal has evolved a response will vary from species to species. Furthermore, more than one proximate factor may be implicated and these may differ with regard to the strength of their biological action.

1.3   Environmental synchronizers

Many kinds of information can be used as proximate synchronizers and more than one kind may be utilized simultaneously. Generally, however, species have tended to evolve responses to those environmental changes which constitute the most stable source of predictive information. At mid and high latitudes the annual variable which is uniquely consistent is the change in daylength, and it is therefore not surprising that many of the animals occupying high latitude habitats appear to use the annual changes in the duration of daylight to orientate themselves and phase the endogenous events leading up to breeding. Thus, the increasing daylengths that occur in early spring stimulate many birds to take up territory, and start nest building and other behavioural activities associated with the release of hormones from the developing gonads. Conversely, some animals such as the Southdown domestic sheep are equally susceptible to decreasing daylengths.

It must be emphasized, however, that many different kinds of environmental information are utilized by different species of animals. Seashore animals, for example, have adapted themselves to the tidal rhythms caused by lunar gravitational influences, and the Palolo worm and its lunar-timed breeding cycle has already been mentioned. In arid parts of the world individual animals may not breed for several years. Then, after heavy rainfall they reproduce repeatedly. The reproductive cycles of some macropod marsupials are thus drought adapted, breeding only when rain falls, and irrespective of the seasonal changes in daylength, and many xerophilous birds are similarly adapted. Again, while seasonal daylength changes have obvious advantages as timing factors in temperate areas, the same is not true for tropical and equatorial environments where the annual fluctuations in the daily light cycle are relatively slight or absent. Here seasonal rains and associated changes in the vegetation act as the proximate environmental synchronizers. In Indonesia, for example, most avian species, including those of evergreen habitats, breed from January to March in west Borneo, and March to July in mid-Java, according to the different time of the rainy season in the two areas.

The above observations indicate some of the geophysical phenomena which have significant effects on living organisms—lunar cycles, tidal cycles, rainfall cycles, and the seasonal changes in daylength and environmental temperatures. They all serve as sources of predictive information which are used by different animals in their varying habitats. In this volume we shall be confining our interest to a study of the effect of the photoperiod as a proximate synchronizer, and other factors will be considered only where it has been shown that they may have a modifying influence on a photoperiodic response. The term *photoperiod* will be used to denote the lighted portion of a natural or experimental cycle and the term *scotophase* will be used to denote the dark period.

1.4   Photoperiodism and photoexperimentation

In most temperate zone, sub-arctic and arctic animals, photoperiodic mechanisms are the major synchronizers. Such controls must have become especially important during the evolutionary conquest of dry land where temperature is much less reliable as an indicator of seasonal changes than in the oceans. That such a correlation exists between daylength and annual cycles has been known in some species of animals for a very long time. For centuries the Japanese have practised 'yogai' which is a method of inducing out of season singing in captured male songbirds by keeping them in artificial long photoperiods in winter. In the Netherlands too this has been practised since very early times.

At the beginning of the century Professor E. A. SCHÄFER suggested on theoretical grounds that daylength might be a determining factor in the migration of birds, and argued that the annual gonad cycles might likewise be controlled by light. But it was not until the nineteen twenties that the experiments of the Canadian zoologist William ROWAN convincingly proved the truth of this theory. Rowan provided the first demonstrable proof that daylength could have an effect on reproductive behaviour in vertebrate animals.

The greater yellowleg is a species of bird which breeds in Canada, migrates to Patagonia in the autumn, and returns again to its Canadian breeding grounds the following spring. This journey involves an overall distance of some 16,000 miles for the round flight, yet the precision of the timing of the migration and breeding is such that the eggs are hatched between May 26th and May 29th each year. Rowan made this observation over a period of fourteen years, and was so impressed by the precision of each annual cycle that he made a thorough analysis of the factors that might explain the astonishing regularity. He considered many factors which vary with the seasons such as temperature, daylength, sunlight, barometric pressure and food, and concluded that only one of these was precise enough to play a synchronizing role: the increasing daylengths after December 21st. To test his hypothesis he trapped slate-coloured juncos, *Junco hyemalis*, a species of bird which winters in Canada, and subjected them to artificially increased daylengths. These artificial photoperiods were gradually increased so that after a few weeks the birds were experiencing daylight conditions that were not normally present until late spring. On examination, control specimens which had been kept under natural winter light conditions, had remained sexually regressed with inactive reproductive organs, but those birds which had been exposed to an unseasonally increased photoperiod were in breeding condition. Having thus discovered that reproductive activity could be manipulated by daylength, Rowan subsequently released birds in various states of reproductive development to note the effect on their migratory behaviour. He concluded from his observations that the birds migrated when their sexual organs were expanding but not when the gonads were inactive or in the full breeding state. For the first time there was experimental evidence that the timing of the spring migration was under the control of the environment.

The work of Rowan is a landmark in the history of the study of photoperiodism. His pioneer investigations triggered off an era of photoexperimentation which has since established the general importance of photoperiodism in the reproductive cycles of a large number of vertebrate and invertebrate species. But although daylength may be the major source of information in the control of many such cycles, it must be remembered that it is probably rarely the sole source, and the extent to which it participates varies extensively from species to species. In some, the seasonal photoperiods seem to have absolute control over the timing of the annual reproductive rhythm, and the white crowned sparrows (*Zonotrichia* species) of North America, for example, will show no signs of migratory activity or gonad development, even at the height of the natural breeding season, if they are retained under an artificial winter daylength. Some species, on the other hand, appear to have an endogenous reproductive rhythm, and the gonads can undergo a cyclic development into full breeding condition even though the animals may be kept in conditions of continuous darkness or unvarying daily photoperiods. For example, when

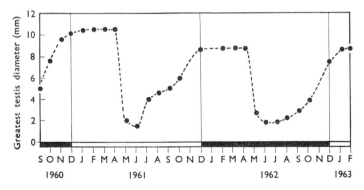

Fig. 1–1   Seasonal changes in the testicular size of the weaver-finch, *Quelea quelea*, kept under an unchanging daily 12-hour photoperiod, (LOFTS, B. (1964). *Nature, Lond.*, **201**, 523.)

equatorial weaver-finches, *Quelea*, were kept under an unchanging daily twelve hour photoperiod for two and a half years, the gonads came up into full breeding condition and regressed with the same periodicity as the gonads of control specimens living in the wild (Fig. 1–1). Internal rhythms such as the latter, are now thought to be fairly widespread among animals, although it is doubtful whether they ever exclusively control the annual cycle. Environmental factors are still necessary to synchronize the periodicity of the rhythm to seasonal events. A. J. MARSHALL has likened such an endogenous reproductive rhythm to a clock in which the cogs seasonally engage various environmental 'teeth' to which the species has evolved a response, and which have the capacity to accelerate or retard the timing. In other words, environmental factors such as photoperiodism serve to 'set' the biological clock.

Photoperiodism in Birds

Among the vertebrate animals the most highly evolved photoperiodic mechanisms occur in birds, and more is known about photoperiodism in these animals than in any of the other vertebrate groups, both from the standpoint of the way in which they measure time, and also how photic information is translated into the complex neuroendocrinological changes associated with the seasonal development of the reproductive organs. Historically too, the causal relationship between long light periods and sexual periodicity has been known in birds for much longer, even before Rowan's photoexperimentation became a subject of major scientific interest. Nearly two centuries ago the Spaniards were subjecting hens to artificial illumination at night to produce an increase in egg production—a technique which is now an established practice in poultry husbandry.

The literature on avian photoperiodism is extensive. It has been established that there are photoperiodic controls in no less than sixty avian species from widely separate orders, and although other secondary factors, such as a sudden cold spell, or an unseasonal period of warm sunny weather may retard or accelerate the rate of gonadal development, there is no experimental evidence that these factors by themselves can initiate gonadal growth in the absence of stimulatory photoperiods. The most precise photoperiodic control mechanisms known in vertebrate animals are those of migratory birds whose breeding and wintering ranges lie within the northern hemisphere.

Photoperiodism and its relative importance in the overall control of gonadal cycles varies extensively from species to species. In some birds, as has already been mentioned, it is absolute, and in the absence of the appropriate photostimulation both breeding and migration are completely suppressed, but others, particularly xerophilous and equatorial species, are less dependent, although even in these species there is evidence that many have retained an ancient capacity to respond to photostimulation when subjected to experimentally prolonged daylengths. The weaver birds, *Quelea*, are examples; when caught in the non-breeding season they can be artificially stimulated into full reproductive condition by long photoperiods. In their natural equatorial habitat, however, daylengths show little annual fluctuation and do not control the reproductive cycle. Rather, it is rainfall and the subsequent regrowth of the pliable green grass which the bird uses when weaving its nest that are the environmental synchronizers to which the bird responds. Without it successful nest-building cannot take place. Thus, although a response to experimentally lengthened photoperiods can be demonstrated, natural selection has favoured the adoption of more appropriate stimuli to control reproduction and ensure its fulfilment at the most propitious period for survival.

2.1 The photoperiod and migration

Migration is a behavioural pattern of great antiquity and was established in birds at least before the last glacial epoch. With their aerial mobility and high metabolic rate birds are well equipped for the development of this phenomenon and it is common to many species. It may be regarded as a behavioural adaptation which brings the migrating individuals to areas more favourable to their breeding activity than they would otherwise have had. By migrating to northern breeding grounds, birds acquire the advantage of longer days and a great, although temporary, abundance of food enabling the successful rearing of young. By migrating south at the end of the breeding season, they avoid the many rigours of winter and, in the case of many northern species, failure to migrate south would mean extinction.

Man has certainly known and wondered about the migration of birds (and other animals) for centuries, but the earlier observations and explanations were based mainly on superstition. It was only at the beginning of the present century that scientific methods of identification and ringing started producing accurate data about bird movements and defined the routes between wintering and breeding areas. The early field work provided much useful data and emphasized the unfailing punctuality and incredible precision which some species display in this cyclic activity.

In the study of avian migration the scientist has been aided by a behavioural phenomenon uniquely displayed by migrants. Normally, birds show little activity at night, but during the migratory period caged individuals develop a marked nocturnal restlessness (*Zugunruhe*). This night activity in caged birds only occurs during the migratory period and never appears in non-migratory species, and it is generally believed to have the same physiological basis as migration. *Zugunruhe* thus provides a measurable quantity with which to test experimentally the effectiveness of various environmental and physiological factors in releasing the migratory urge in different species, and is a most valuable tool for laboratory studies.

Activity patterns are generally recorded by means of a sprung perch attached to a microswitch incorporated into an electric circuit. The springing is adjusted so that when the bird hops on the perch, its weight depresses it sufficiently to close the microswitch and circuit, and an electrical impulse passes into a recorder. By this means the number of impulses (i.e. hops) per unit of time is recorded and the diurnal activity pattern assessed. The activity of a non-migratory species over a 48-hour period, together with that of a migratory white-crowned sparrow (*Zonotrichia leucophrys*) during and after the migratory season is shown in Fig. 2–1. It can be seen that the sedentary nutcracker (*Nicifraga columbiana*) and the white-crowned sparrow during the non-migratory period, both have their activity confined to the daytime, but in the migratory phase the latter bird has well marked nocturnal activity as well. This has been shown to be true of a large number of migratory species.

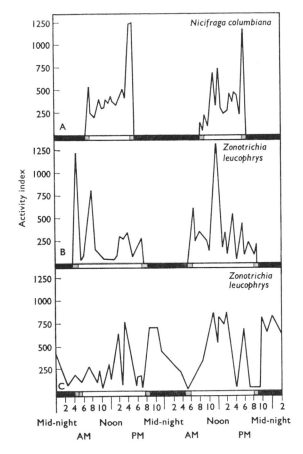

Fig. 2–1 48-hour records of perch-hopping activity in the non-migratory nutcracker (*Nicifraga columbiana*) in November (A), and in the migratory white-crowned sparrow (*Zonotrichia leucophrys*) during the non-migratory season (B), and during the migratory period (C). (FARNER, D. S. and MEWALDT, L. R. (1953). *Bird-Banding*, **24**, 55.)

During the course of a year migratory species display two *Zugunruhe* peaks coinciding with the pre-nuptial journey to the breeding grounds and the post-nuptial return to the wintering areas. Birds taken on their wintering grounds do not show nocturnal activity. The brambling, *Fringilla montifringilla*, is a winter visitor to Britain which migrates north to its Scandinavian breeding area in April. *Zugunruhe* has been observed to start developing in caged specimens early in April, and reaches a maximum intensity by the end of the month which coincides with the main migratory wave of the wild population. In winter time no *Zugunruhe* is recorded, and if bramblings caught during January and February are maintained on an artificial 8-hour photoperiod throughout March, April and May, migratory restlessness fails to develop, even though the wild population has meanwhile migrated north. If, at the end of this period the light is then adjusted to give the equivalent of a mid-April photoperiod, *Zugunruhe* activity is rapidly developed. The results of such an experiment are shown in Fig. 2–2 and they indicate the very significant influence that the photoperiod has on this behavioural activity.

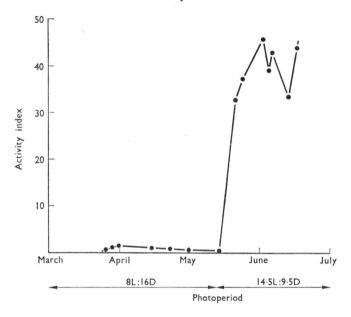

Fig. 2–2   *Zugunruhe* activity in caged bramblings. When the birds were kept under an 8-hour photoperiod throughout the migratory season nocturnal restlessness never developed, but as soon as the daily photoperiod was readjusted to 14·5 hours, *Zugunruhe* activity became established within a few days. (LOFTS, B. and MARSHALL, A. J. (1960). *Ibis*, **102**, 209.)

Similar correlations have been observed in a number of avian species and provide suggestive evidence that the annual timing of the spring migration is photoperiodically activated. Premature vernal *Zugunruhe* can always be induced by keeping wintering migrants under artificial spring photoperiods. The factors which stimulate the post-nuptial autumnal migration, however, are less well understood since it occurs at a time when the natural photoperiod is decreasing and environmental temperatures are dropping. Furthermore, many birds are in a photorefractory state during this period (see Chapters 2 and 3).

The physiological mechanisms by which the photoperiod mediates its effect and initiates the migration are unknown although an indirect control via the secretion of gonadal hormones has often been postulated. As early as 1824 JENNER suggested that gonadal enlargement was the prime cause that excited birds to migrate, and the later photoexperimentation of Rowan was generally believed to substantiate these conclusions. Many later investigators also supported this gonadal hypothesis and held that spring migration was initiated by gonadal hormones. *Zugunruhe* activity, however, is independent of gonadal activity even though this migratory unrest develops parallel to the spring recrudescence of gametogenesis in the gonads of the migrating birds. It is now clearly established in a number of migratory species that *Zugunruhe*, of equal intensity to controls, develops in castrated species, and is thus independent of endogenous sexual stimuli.

Before they start their long journey, migratory birds rapidly increase in weight due to the laying down of large deposits of subcutaneous and visceral fat reserves used as food reserves *en route*. These heavy deposits of pre-migratory fat are generally laid down in spring within ten to twelve days. This accumulation, however, is not necessarily an essential prerequisite for migration, and *Zugunruhe* can develop in birds which have been experimentally prevented from laying down large quantities of depot fat by strict dietary control. There is a similar period of migratory fattening in the late summer, except that the rate of gain appears to be slower than in the spring. This phenomenon is controlled by photoperiodic mechanisms and the laying down of vernal fat reserves can be artificially advanced by changing the daily photoperiod from a short 8-hour daylength to a long 16-hour daylength. Thus, both spring *Zugunruhe* activity and migratory fat storage, though independent phenomena, are controlled by mechanisms using the same primary source of photoperiodic information.

2.2   The photoperiod and gonad response

The gonads of birds display very marked morphological variation throughout the year. The change in size is quite remarkable and can produce up to a 500-fold increase in weight. The annual cycle in testicular size for a number of species is shown in Fig. 2–3 and will serve to indicate the general patterns. The gonads are generally at their minimum size during the winter months when they are gametogenetically inactive, then they start expanding in the early spring. The rate of growth accelerates with further increases in daylength until the maximum size is attained at the appropriate time of the year. The morphological change reflects the resumption and build-up of gametogenesis until, in the case of the male, spermatozoa are being produced in large quantities. The histological changes which accompany this growth in size are shown in Plate 1. The production of sex hormones by the interstitial endocrine tissue also increases parallel to gametogenesis, and stimulates the development of the

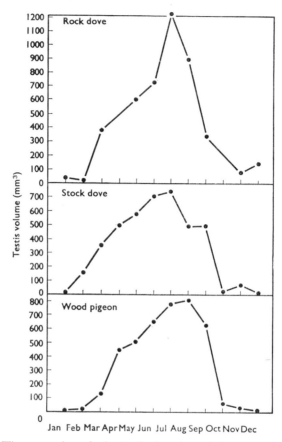

Fig. 2-3 The annual cycle in testicular size of three species of British birds. (LOFTS, B., MURTON, R. K. and WESTWOOD, N. J. (1966). *J. Zool. Lond.*, **150**, 249.)

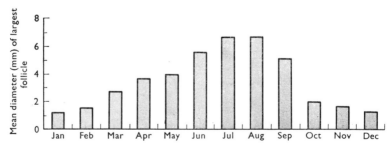

Fig. 2-4 Seasonal variation in the mean diameter of the largest ovarian follicle in the wood-pigeon. A follicle diameter of over 5 mm indicates an ovary close to egg-laying.

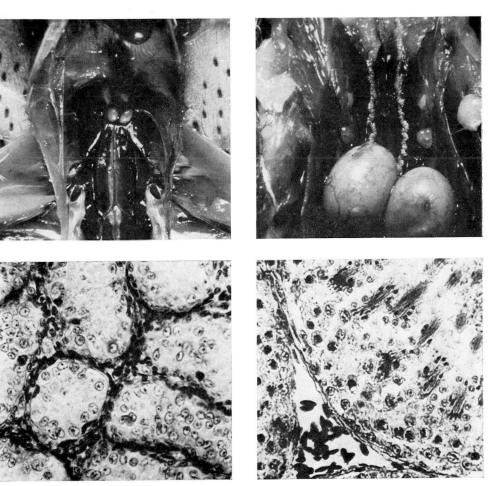

Plate 1 The upper photographs show the tremendous contrast in the size of the testes and accessory sperm ducts, in a tree sparrow in winter and at the height of the breeding season. The two lower photographs show the spermatogenetic condition of the equivalent stages. In winter, the seminiferous tubules are small and contain only a peripheral ring of spermatogonia, whereas the seminiferous tubules of the breeding testes are large and filled with spermatozoa.

(a)

(b)

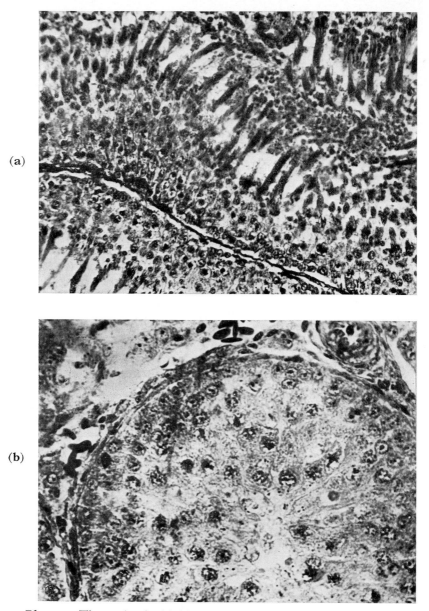

Plate 2 The testis of a bird kept under a highly stimulatory photoperiod (a) has expanded seminiferous tubules filled with bunches of spermatozoa. Reduction of the daily photoperiod to only 6 hours (b) causes testicular regression and within a few days sperms disappear and degenerate germ cells can be seen in the tubules.

accessory ducts as well as the associated behaviour patterns of taking up territory, courtship behaviour, culminating in copulation and fertilization of the egg. The annual cycle in ovarian weight follows the same general pattern as that of the testis. The period of reproduction terminates with the rapid regression of the gonad at the end of the breeding season (Fig. 2–4).

The relative importance of daylength for starting spring gonadal activity varies extensively from species to species. In many the annual gonadal development is rigidly dependent upon the return of long daylengths in the spring. Such birds experimentally held on a winter photoperiod of 8 hours will never develop into breeding condition, whereas a summer photoperiod of 16 hours given in mid-winter will accelerate them into full gametogenetic activity. By experimental manipulation of artificial light regimes to give a session of long photoperiods followed by a few weeks of winter daylengths, and a return to a further phase of long daylengths, such birds can often be induced to develop two breeding cycles within a single year. The brambling is an example of such a species. On the other hand, some birds have an autonomous rhythm of gonadal development and when they are kept under conditions of total darkness, the gonads still undergo a cyclic enlargement and regression. This has been shown to be the case in domestic ducks and drakes, and is also true of many xerophilous species e.g. zebra finches. In such species light is not obligatory for initiating spermatogenesis, but when the birds are isolated from natural seasonal photoperiods and kept under artificially constant conditions for long periods, the endogenous cycles become irregular both in amplitude and frequency. Eventually, the endogenous cycles becomes out of phase with the normal seasonal events, so that here too environmental photofluctuations are essential to synchronize the internal rhythm with the optimal seasonal conditions.

In some circumstances there is evidence of temperate birds coming into breeding condition regardless of the seasonal photoperiod, but this is usually a secondary adaptation resulting from domestication and the consequent provision of a continuous abundance of food. Such is the case in the flocks of feral pigeons so common in our towns and dockyards. The town pigeon is a descendant of the wild rock dove (*Columba livia*) which inhabits sea caves and is a strictly seasonal breeder with an obligatory photoperiodically controlled gonad cycle. Domestic strains of this species have been kept by man since early Egyptian times and have been maintained in dove-cotes in Britain since early Norman times. Escapes from various domesticated sources have given rise to the now free-living feral populations which have an all-the-year-round supply of food from a benevolent public or through spillage at the docks. The elimination of a seasonal food supply has permitted the abandonment of the normally precise annual cycle and many town individuals now breed throughout the year. An example of the seasonal variations in follicle size of one such

2*

population is given in Fig. 2-5 and can be compared with Fig. 2-4. A mean
follicle diameter of over 5 mm indicates a population close to egg-laying.
A similar phenomenon has also become established in the domestic hen.
In the course of selection for high egg production, poultry breeders have
developed strains and breeds in which a restricted seasonal breeding pattern
has been eliminated. Natural photoperiods no longer stimulate a vernal
ovarian recrudescence but simply regulate the rate of egg production from
a permanently developed ovary. Under natural photoperiods the maximum
rate of egg-laying occurs during the longest days and declines in autumn
and winter, but nowadays it is common practice to use artificial lighting in
the poultry houses to stimulate egg production when it is normally low.

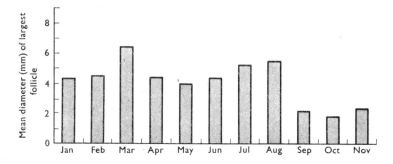

Fig. 2-5   Seasonal variation in the mean diameter of the largest ovarian
follicle in a population of feral town pigeons. The seasonal trend noted in the
wood-pigeon (Fig. 2-4) has been eliminated. (LOFTS, B., MURTON, R. K. and
WESTWOOD, N. J. (1966). *J. Zool. Lond.*, **150**, 249.)

Under experimental conditions the testes of probably all avian species
can normally be brought to a state of complete spermatogenesis by light
stimulation. Fig. 2-6 shows the testicular growth curve for the Japanese
quail *Coturnix c. coturnix* when they are kept under the highly stimulating
condition of a 20-hour daily photoperiod. In this species the rate of growth
is particularly great, and the testes increase in weight from 8 mg to 3000 mg
within three weeks of exposure to these conditions. The importance of
light is emphasized by the rapid testicular regression induced by switching
the birds to short daily photoperiods. Histologically, the testes show signs
of degeneration within a few days of reducing the photoperiod (Plate 2).

In female birds, although they display a similar gonadotropic response
to a photic stimulation and are undoubtedly synchronized in the wild by
the natural photoperiods, under laboratory conditions the ovary can rarely
be brought up to full breeding dimensions, a block occurring at the onset
of vitellogenesis. To develop beyond this point, appropriate psychic
stimuli, such as the appearance of a breeding mate and engaging in inter-
pair courtship behaviour and other such phenomena, are a necessary pre-

requisite before the final phase of ovarian growth can be achieved. Captivity itself is sufficient to block the ovarian development. Thus, female white-crowned sparrows caught during their spring migration, undergo very little further gonadal development when maintained in large outdoor aviaries, whereas the male birds progress into full sexual maturity. An exception appears to be *Coturnix*, where young females under long photoperiods, develop to full reproductive condition.

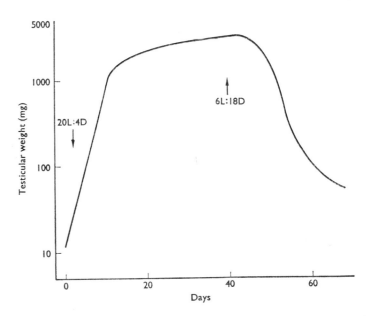

Fig. 2–6   Testicular growth curve of Japanese quail exposed to a 20-hour daily photoperiod. After 40 days the light cycle was readjusted to give a 6-hour photoperiod. (LOFTS, B., FOLLETT, B. K. and MURTON, R. K. (1970). *Mem. Soc. Endocr.*, **18**, 545.)

The testicular growth induced by stimulatory daily photoperiods of constant duration is approximately a logarithmic function of time. The rate of growth can thus be represented by the following formula:

$$k = (\log W_t - \log W_0)/t$$

where W_0 is the resting testicular weight, W_t is the testicular weight after t days of photostimulation, and k provides us with a quantitative measure of the rate of gonadal growth. It has been established in a number of species that k is dependent on the daily photoperiod and bears a linear relationship to the duration of the photoperiod over a wide range of values (c. 9 to 18 hours). The value for *k* differs for different species and provides an objective basis for valid comparisons between the effectiveness of different photoperiodic treatments in different birds (Fig. 2–7).

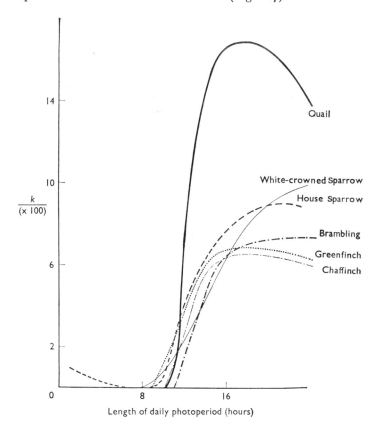

Fig. 2–7   Rate of testicular development (*k*) as a function of the daily photoperiod, in six different species. Note the linear relationship over the photoperiod range for about 9 to 18 hours. (LOFTS, B., FOLLETT, B. K. and MURTON, R. K. (1970). *Mem. Soc. Endocr.*, **18**, 545.)

That gonad stimulation varies with the length of the photoperiod is evident from an analysis of the data on maximum gonad sizes achieved in wild populations of the same species living at different latitudes. Daylengths are longer the higher the latitude, and generally, in populations of the same species distributed over a large range, the more northerly populations, living under much longer daily photoperiods, possess gonads which develop to a significantly larger size than those of the birds occupying habitats in lower latitudes. Examples of this correlation between gonad size and

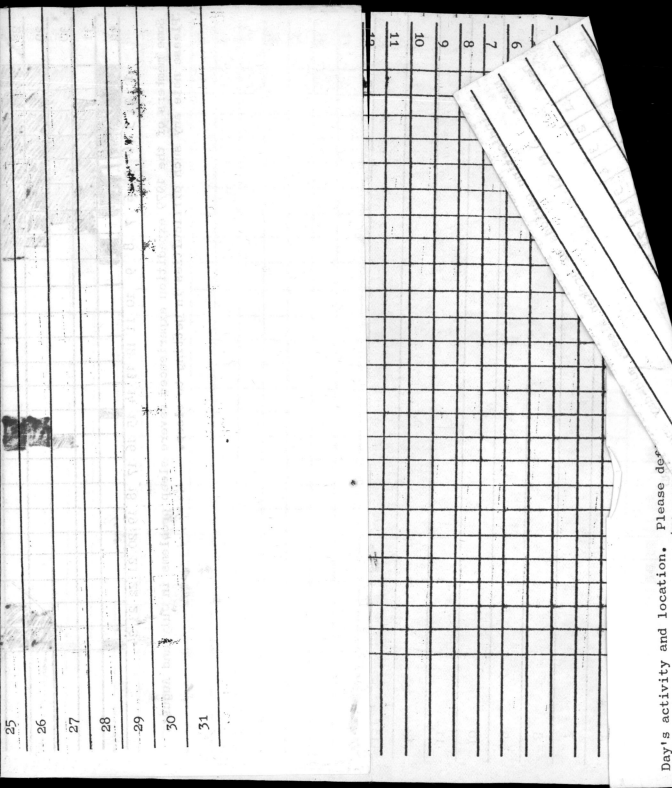

Day's activity and location. Please des...
note position by grid-square (Figure

Date

25
26
27
28
29
30
31

latitude (and hence photoperiod) are given in Table 1. For most species there is a daily photoperiod below which no gonad growth occurs, and also one which induces the maximum rate of development.

Table 1   Maximum testis volume of populations of the same species at different latitudes.

Species	Latitude	Maximum gonad size (cu mm)
Passer domesticus	34	400
	45	490
Anas platyrhynchos	47	700
	52	1500
Columba palumbus	52	450
	57	955

As has already been briefly noted elsewhere, the gonadal response to stimulatory photoperiods may be modified by a host of other environmental variables (temperature, weather, availability of a nesting site, etc.), which may be mutually antagonistic, and can be broadly classified into accelerators or inhibitors. They do not initiate the spring gametogenetic recrudescence, this can only be done by the appropriate stimulatory daylength, but they are important co-ordinators timing the final oviposition. Temperature is perhaps the most important factor which may modify the cycle in this way, and there is much field data showing the correlation between temperature and the speed of testicular development, and also the date of egg-laying. Thus, it is well documented that an exceptionally mild spring will cause unusually early breeding in many resident passerine species, whereas, conversely, unusually cold weather at the same time of year can nullify the effects of spring daylengths and retard, or if sufficiently severe, temporarily stop, gametogenetic development. In particularly mild winters, resident British species such as the blackbird, song thrush, robin, starling, house sparrow and skylark, sometimes begin laying eggs in late November and early December. Normally, although all these birds show some upsurge of sexual activity in the autumn, they do not usually become sufficiently sexually advanced to lay eggs, and no further physiological development occurs until the following spring. It is possible, however, that for all these species the photoperiod is sufficiently stimulatory in the autumn, but is normally prevented from evoking a gonadal development by the low ambient temperature. Only when exceptionally mild environmental temperatures are encountered, is a resurgence of gametogenesis possible. Among water-birds and many others, a frequent inhibitor is the lack of a safe and traditional nesting site. If the acquisition of such a site is denied for too long it can lead to a regression of gonadal development, even though the species may be experiencing highly stimulatory photoperiods.

2.3   The refractory period

In many photoperiodic species the gonads spontaneously regress at the end of the breeding period. This comes immediately after reproduction in single-brooded birds and after the final ovulation of the season in multi-brooded ones, and usually takes place at a time when the bird is still experiencing normally stimulating photoperiods. Once the birds have entered this stage they no longer respond to photostimulation, even when subjected to an experimental light period in excess of natural summer day-lengths. They have become photorefractory and are said to have entered the *refractory period* of the annual reproductive cycle. Since they now no longer respond to an environmental stimulator, the refractory period imposes a temporary break in breeding potential and obviously has a very important synchronizing role. The duration for which a bird remains photorefractory is variable, being as short as four weeks in some species and as long as four months in others. Many early breeding species have refractory periods of the latter type, thus the spring breeding mallards, *Anas platyrhynchos*, soon terminate gametogenesis and become photo-refractory until the autumn. When the refractory period is over they are once again capable of being stimulated by long photoperiods, but remain sexually regressed due to the inhibitory factors such as cold temperatures and diminishing daylengths which are then prevalent in their environment.

The development of photorefractoriness may be regarded as an adaptive safety mechanism, preventing the production of young when those environmental factors necessary for successful rearing are becoming reduced, but daily photoperiods are still long. In migratory species it ensures that breeding terminates sufficiently early to enable the post-nuptial moult to take place and the new feathers to be regenerated in time for the return migration. Also, the birds can lay down their migratory fat at a time when there is plenty of food available. Similarly, juveniles need time to develop before they migrate, and late-hatched birds would be at a selective disadvantage in this respect. Usually, birds which have to undertake long migrations enter their refractory phase earlier than those which undertake shorter journeys. In sedentary species it ensures that clutches are not produced too late in the season and that nestlings are not hatched out when food is becoming less abundant. For example, the rook, *Corvus frugilegus*, is stimulated to a reproductive state by late February daylengths so that young are produced in April and May. It then enters its refractory period which persists until the autumn. If this were not so, the highly stimulatory photoperiods of June to September would maintain the reproductive potential at a time when earthworms, an important food supply, are hard to collect because they burrow deeper in the dry conditions of summer. Thus heavy juvenile mortality would be the consequence of reproductive activity at this time. The onset of refractoriness acts as a safety mechanism until the period of long and stimulating daylengths is over.

Photorefractoriness is very widespread among avian species, particularly among the arctic, sub-arctic and temperate zone migrants. Experimental analysis has shown that it may also occur even among some equatorial birds as an integral part of their autonomous rhythm. The endogenous cyclic enlargement of the testis in *Quelea* under an unvarying 12L : 12D (i.e. 12 hours light and 12 hours dark) regime, discussed in the earlier part of the book (Fig. 1–1), is followed by a spontaneous regression, and birds exposed to 17-hour photoperiods during this period are found to be photorefractory. They remain so for a matter of 6 weeks.

It is perhaps relevant to mention at this point that a refractory stage is also known to occur in the reproductive cycles of some fishes, reptiles and mammals. It is a phase of the reproductive cycle which has, like any other part, been exposed to selective pressures. Generally, in opportunist breeders like many xerophilous birds that respond to sporadic rainfall, or species which have become highly domesticated, the duration of the refractory period has become greatly abbreviated so that the reproductive potential is rapidly regained and a second breeding cycle can be induced to exploit the favourable conditions which may be available for only a very limited period or, in the case of domestic animals, to exploit conditions which by artificial means are now available all the year and not just seasonally. An example of such an adaptation is seen in the budgerigar which, after experiencing sudden heavy rainfall in its normally arid environment, reproduces repeatedly so that adults and new young become so numerous that they breed in hollow sticks that have fallen from the trees to the ground. Here the refractory period has become greatly foreshortened and there is only the briefest interruption of the reproductive potential.

Photoperiods can also affect the duration of the refractory period and, in many species, reduced photoperiods are necessary to terminate the phase and enable the bird to retain its photosensitivity. In a number of species the termination of photorefractoriness can be experimentally advanced by artificially exposing them to a period of short days and, conversely, in many photoadapted birds continued exposure to long daylengths during the refractory phase will prevent an emergence from this part of the cycle so that they do not regain a reproductive condition. It is only after exposure to a species-specific period of short daylengths that these animals regain their capacity to respond to increasing daylengths. If, for instance, refractory male mallards are placed under an experimental 16-hour daylength, the testes will remain regressed until the birds experience a period of short winter daylengths. Eight weeks in an 8-hour photoperiod is sufficient to break the photorefractoriness. In species like *Quelea*, however, which are no longer photoperiodically synchronized yet still retain a residual, albeit greatly abbreviated, refractory phase, prolonged exposure to long photoperiods does not prolong the refractory period and birds spontaneously emerge after a species-specific period of time, and rapidly regain their reproductive potential. Thus, the dependence on experiencing a period of short days to eliminate photorefractoriness is absolute in some species like the wild mallard, juncos, white crowned sparrows and starlings, but others, like *Quelea* and the domestic duck, can regain full testicular development whilst being held under a continuous light regime.

The effect of light on the refractory period is very clearly demonstrated in the house sparrow, *Passer domesticus*, which has a range extending through many parts of the world following introductions by man. Data on the testicular cycle of different populations are shown in Table 2. From

Table 2   Duration of active spermatogenesis in different populations of *Passer domesticus*.

	Latitude	Approximate photoperiod at summer solstice (hours)	Duration of active sperm production (days)	Length of refactoriness (days)
Pasadena	34	$14\frac{1}{2}$	138	13
Norman	36	$14\frac{1}{2}$	135	31
London	43	$15\frac{1}{2}$	118	64
Belfast	54	17	118	47

these it can be seen that the more stimulating the photoperiod, the sooner the bird enters its refractory phase. Thus the two lower latitude populations in Pasadena and Norman which achieve full spermatogenesis by mid-February and early March respectively, finish sperm production and enter a refractory state two weeks to a month later than more northerly populations exposed to more stimulatory photoperiods. Furthermore, the duration of the refractory period tends to be shorter in the southernmost populations which experience the shorter summer daylengths.

In general, southern hemisphere birds have the same photoperiodic mechanisms as their northern hemisphere counterparts, i.e. the gonadal cycles are timed so that the gonads recrudesce under increasing spring daylengths. In cases where such species have been transported across the equator into northern latitudes, their reproductive cycles have become readapted and synchronized by the new seasonal photoperiods and breeding takes place in the northern spring and summer. Long distance trans-equatorial migrants such as the bobolink, *Dolichonyx oryzivorus*, and short-tailed shearwater, *Puffinus tenuirostris*, breed in one hemisphere and winter in another so that they experience two light cycles while undergoing one reproductive cycle. The bobolink for example, leaves its North American breeding ground in late August and has migrated beyond the southern shores of the United States by mid-October. During this southern migration the birds are experiencing a rapidly diminishing daylength, but once they cross the equator and reach their South American wintering grounds the photoperiods are once again increasing. In short, after their June breeding season in North America, the birds experience a period of short daylengths on their southward journey and then a stimulatory photoperiod in the southern hemisphere, yet they do not breed there. These birds show an adaptation of their refractory period which is unusually long compared with temperate zone migrants, some eight and a half months (mallards have a 4-month photorefractory phase). This ensures that wintering bobolinks are insensitive to the long photoperiods experienced in their wintering area and they only regain their photosensitivity after the southern solstice when daylengths are decreasing and they start their northward nuptial flight. Once past the equator they then respond to the northern spring light conditions.

A characteristic period of post-nuptial refractoriness is not an inevitable part of all avian breeding cycles, and in at least one species where it would have no adaptive value photoexperimentation has shown it to be absent. In the case of the wood-pigeon, a refractory period is unnecessary, as it is a resident species which can take advantage of an autumnal grain harvest to provide for late-hatched young. When these animals are subjected to stimulatory daylengths at the end of the breeding season, gametogenesis is maintained in those individuals which are still in breeding condition, while birds which have already become sexually regressed are immediately stimulated back into full spermatogenesis. In contrast, a photorefractory state is an essential component of the closely related, but migratory, turtle dove, *Streptopelia turtur*, which migrates to central Africa for the winter, and therefore requires a safety device to curtail breeding and allow sufficient time for the necessary pre-migratory changes to occur. In this species, gonadal regression will occur even if the birds are held on long photoperiods. The refractory period sets in during July and females killed during the first week in August show no follicles in excess of 1 mm diameter. If these birds are placed under a 17-hour photoperiod, they still show no sexual advancement when killed for autopsy six months later.

2.4   Interspecific and intraspecific differences in photosensitivity

It is clear from the wide diversity of breeding patterns displayed by different species that both intraspecific and interspecific differences in the photoperiodic response must exist. This, of course, is not surprising in view of the fact that natural selection must operate on *all* of the physiological mechanisms controlling the reproductive cycles. In photoperiodically controlled species where the neuro-endocrine mechanisms regulating reproductive development are synchronized by light, one might anticipate that in closely related groups breeding at different times of the year, selective pressures have led to the emergence of differences in photosensitivity which result in the variations in reproductive patterns. Such a phenomenon is very well exemplified by the North American *Zonotrichia* populations which include several species and sub-species forming a complex of geographical variants over the Continent. These birds have been extensively used for photoexperimentation and there is a great deal of data about the gonadal response under varying light conditions. In California, there are three races of the white-crowned sparrow, *Zonotrichia leucophrys*. These are, *Z.l. pugetensis* which is a summer visitor in many parts of the state; *Z.l. nuttalli*, which is a permanent resident within a restricted range; and *Z.l. gambelii*, which is a winter visitor from mid-September to late April. Two other closely related species, the golden-crowned sparrow *Z. atricapilla*, and white-throated sparrow *Z. albicollis*, also occur in winter. The migratory species leave California and fly north to their breeding grounds in the spring; Fig. 2–8 shows the distribution of the breeding grounds of these various birds. From this it can be seen that the breeding area of *Z.l. nuttalli* being a non-migrant has a restricted

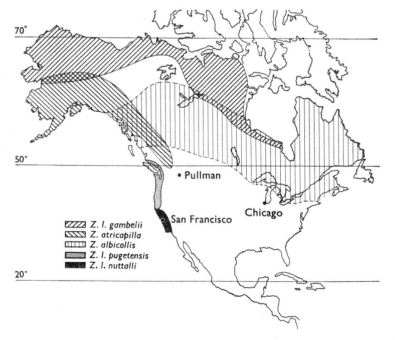

Fig. 2–8   Breeding grounds of *Zonotrichia* species and races. (LOFTS, B. and MURTON, R. K. (1968). *J. Zool. Lond.*, **155**, 327.)

range in California, whereas *Z. albicollis* and *Z.l. gambelii* migrate north to breed in June, the former occupying the scrub niche in the northern conifer forests, and the latter in the Canadian arctic, to the west and south-west of Hudson Bay where they breed in the deciduous thicket of the low tundra. *Z.l. leucophrys*, also migrates to the arctic areas, occupying the regions to the east and south of Hudson Bay. The nuptial grounds of *Z. atricapilla* are the sub-alpine alder and willow thickets of the mountains of British Columbia and Alaska, where breeding takes place in late June and July. The migratory range of *Z.l. pugetensis*, by contrast, is much shorter and breeding takes place along the western seaboard of America.

The experimental data on the photoresponses of the various species and sub-species show that generally, the level of photosensitivity bears an inverse relationship with the migratory pattern, being lowest in the species with the longest northward migration. Thus, the resident Californian species, *Z.l. nuttalli*, has a high level of photosensitivity, i.e. it responds to shorter photoperiods. Therefore in terms of the annual cycle, gonadal recrudescence occurs early in the year and gametogenetic activity continues for a longer period, thus enabling the birds to exploit the long season of optimal conditions in its Californian habitat. This species is multibrooded. In contrast, the migrant single-brooded races which breed in the arctic or northern mountain regions, have a low level of photosensitivity and require much longer daylengths before gonadal recrudescence becomes initiated. As a consequence, full reproductive condition lasts for a shorter time and terminates before the onset of the arctic winter. It is only when they reach the much longer photoperiods of the higher latitudes that these birds are sufficiently photostimulated to develop full breeding condition. The seasonal gonad cycles are shown in Fig. 2–9 which clearly

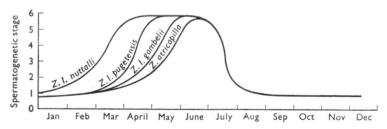

Fig. 2–9   Testicular cycles of four species and sub-species of *Zonotrichia*. The scale 1 to 6 represents the spermatogenetic condition reached by the gonad (1 = spermatogonia only; 6 = bunched spermatozoa). (LOFTS, B. and MURTON, R. K. (1968). *J. Zool. Lond.*, **155**, 327.)

indicates this adaptive aspect of their photoresponses. The photosensitive threshold of any particular species will proximately dictate the pattern of the annual breeding cycle.

Differences in photosensitivity can also be shown by closely related resident species, particularly those that may occupy the same area. This has been clearly established for three closely related species of British pigeons, the wood-pigeon, *Columba palumbus*, stock dove, *C. oenas*, and wild rock dove, *C. livia*. Under natural conditions the gonads of the woodpigeon in Cambridgeshire start their normal vernal gametogenetic recovery in early March, and most adults do not acquire full breeding potential until May. In the same area, the stock dove, on the other hand, begins its recrudescence in January, so that most birds are in full breeding condition by March. This difference in breeding pattern can be accounted for by a species difference in photosensitivity and can be shown experimentally. Spermatogenesis can be induced in the sexually regressed winter stock dove with artificial photoperiods increasing from 9·1 to 10·8 hours at the rate of 15 mins every 3 days, the equivalent of natural daylengths at 52°N in late January to mid-February. Sexually regressed wood-pigeons kept under identical experimental conditions, on the other hand, do not show any gametogenetic response (Plate 3) and these birds only start showing a testicular development when they are exposed to a photoperiod of not less than 11 hours. The adaptive significance of these two quite distinct annual cycles becomes very evident when one considers their food requirements. Stock doves are weed-seed specialists and can rear their young more successfully earlier in the season than the wood-pigeon, which mostly breeds later in the year in relation to the availability of cereal crops.

2.5   The photoperiod and moult

The number of moults and their timing differ from species to species, but generally, the main moult follows the breeding season. In many birds there is often a spring moult as a consequence of which the bird dons a nuptial plumage, but this is usually only a partial replacement of feathers and does not generally involve the renewal of the entire plumage, a characteristic of the post-nuptial moult. The burden on the bird's protein reserves of replacing the feathers is very heavy, and is generally too much to accomplish at the same time as breeding. Only in situations of abundant food supplies are the two processes seen to occur simultaneously, e.g. in wood-pigeons. More commonly, selection has favoured one season for moulting and one for breeding. The need for a quick breeding season followed by quick moult is most urgent in those migrants to high latitudes, since the replacement of the wing feathers must occur sufficiently early to enable the birds to undertake the southerly migration, and must also be achieved before the onset of poor autumn or winter feeding conditions limit the intake of food.

In areas where environmental photoperiodic fluctuations are marked the moult may be dependent on seasonal daylength changes, and there is much evidence for some species that the moult cycles are photoperiodically controlled, to some extent at least, independently of breeding periodicity. Thus, by suitable daylength manipulations, juncos and white-crowned sparrows can be forced to undergo two moult cycles in a year. Reducing the photoperiod advances the time of the post-nuptial moult, whereas long photoperiods during the refractory period tend to retard it. For example, house sparrows caught during the breeding period in July, and held on a 16-hour photoperiod, show a delay in their postnuptial moult so that when they are examined three months later and compared with specimens which have been kept under natural light conditions, the absence of the newly grown white-tipped breast feathers seen in the latter specimens, is very obvious (Plate 4). Wild mallards can be similarly manipulated, and when birds caught during August are placed under 16-hour photoperiods, they retain their brown henny plumage for several months, whereas controls in outdoor aviaries or birds placed under 8-hour photoperiods, moult into the colourful cock plumage within two months. Hence, in at least some north temperate species, a post-nuptial exposure to a period of reduced environmental photoperiods is not only necessary for the birds to regain their photosensitivity but also for the completion of the moult.

Photoperiodism in Mammals 3

Much less attention has been paid to the photoperiodic mechanisms of mammals, and the relevant literature is extremely sparse in comparison with the recorded data on this phenomenon in birds. Whereas in some avian photoperiodic mechanisms the annual changes in daylength are apparently entirely necessary for some cyclic events, such as the release of migratory behaviour, the termination of the photorefractory condition, or the spring gonadal recovery, in mammals the photoperiodic controls are far less rigid and generally serve to time well-developed endogenous rhythms which would result in the eventual development into a reproductive condition, even when the animal is experimentally isolated from its normal environmental cues. In contrast to the pattern of photoexperimentation in birds, much of our knowledge concerning the light-regulated reproductive cycles in mammals has been derived from studying its effects on ovarian and oestrous cycles, and relatively little data exists on how it affects testicular cycles. Another point of difference between birds and mammals is that, whereas in the former group there is no evidence that *decreasing* daylengths activate or accelerate the sexual development of any species so far investigated, in mammals the photoresponses can be both of long-day types and short-day types.

3.1 Long-day photoresponses

There are many wild mammalian species, particularly those that occupy the higher latitudes, which, like the birds, breed in response to an increasing spring photoperiod. In such animals oestrus can be prematurely induced from the sexually quiescent anoestrous state by subjection to prolonged photoperiods and, conversely, retarded by keeping them under reduced photoperiods. An example of this situation is seen in the ferret. Under natural conditions this animal has a long anoestrous period lasting from August or September until the following March, and in the absence of any experimental manipulation by light stimulation or hormone therapy, ferrets have never been known to breed during this period in northern latitudes. However, when male or female ferrets are subjected to prolonged photoperiods in mid-winter they can be stimulated into complete spermatogenesis, and full oestrus, so that breeding takes place and parturition can occur in January. When ferrets are kept under short photoperiods, on the other hand, the development of oestrus is delayed, though this is not an absolute control since it will eventually develop even under these reduced photoperiods, and blinded animals can still come into an oestrous condition.

Light stimulation produces a similar response upon the reproductive activity of the mare. Thus, anoestrous mares subjected to prolonged artificial photoperiods in January can show a resumption of ovarian follicular development within 15 days of the start of such treatment. Furthermore, if these horses are placed under stimulatory daylengths in August, the period when they start entering their anoestrous state, the ovaries continue to be functional through the normally anoestrous period. Like the ferret, the cycle is not completely controlled by light, however, and animals maintained in darkened stables will show a delay in the resumption of ovarian activity but not a complete inhibition. A similar photostimulatory response has been shown in the cat, hare, hedgehog and white-footed mouse, and is probably a synchronizing factor in the seasonal regulation of the reproductive cycles of most spring-breeding mammalian species. The raccoon is particularly susceptible, and can be induced to produce two litters in one year by suitable artificial photomanipulation.

It is true, of course, that, as in birds, not all mammals show the same level of sensitivity to seasonal photofluctuations, and some are apparently independent of seasonal light changes so far as their reproductive rhythms are concerned. The breeding rates of the Orkney vole (*Microtus orcadensis*), for example, does not significantly differ under different experimental light regimes, but the ordinary field vole (*M. agrestis*) is photosensitive and its oestrous cycle can be modified by varying the photoperiod. The guinea-pig is another mammal which appears relatively independent of seasonal photofluctuations, but this is perhaps not too surprising since, in its natural tropical habitat, seasonal light changes are absent or only slight.

As with domestic birds, domestic mammals have shown a similar trend in the loss of dependence on photoperiodism, and many have evolved into continuous breeders in the absence of exposure to the selective influence of a seasonally limited food supply. The domesticated rabbit has lost its pattern of seasonal breeding and is capable of breeding at any time of the year, but in the wild it is still strictly seasonal and can be shown experimentally to respond to additional daily rations of light by a premature appearance of the spring gonadal recrudescence. Even with domestication, however, mating and conception occur more frequently during March to July, then at other times, and the same is true in cattle where there is a strong tendency to calve more frequently in the months of February to April.

3.2 Short-day photoresponses

The sexual development of short-day forms starts at the end of summer and early autumn, and many breeds of sheep, goats and deer are known to be of this type. Because of their economic value much attention has been focused on the control of breeding by light in these species, and the stimulatory effects of decreasing daylengths have been well established. Experimentally, in sheep the sexual season can be completely reversed by artificially providing additional amounts of light during the winter, and decreasing the photoperiods during the summer. The natural period of anoestrus can always be prematurely terminated, and the animal brought into a sexual condition, by reducing the summer photoperiod. The selective processes of domestication have produced demonstrable differences in the photoresponses of the different breeds, but they nearly all become stimulated into breeding condition by declining autumnal days. In England, decreasing daylengths after the summer solstice initiate gonadal recrudescence which result in Suffolk ewes breeding some three to four months later. Their breeding activity finishes when the natural photoperiods start to increase again. The production of spermatozoa by the ram also varies inversely to the photoperiod and similarly encourages successful matings in the late autumn. As a consequence, the lambs are born in the following spring when conditions are most favourable for their survival.

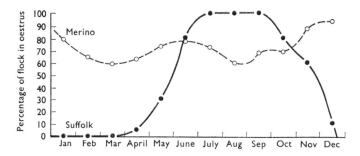

Fig. 3-1   The breeding season of Suffolk ewes in Cambridgeshire, and Merino ewes in Kenya.

Out of season sexual behaviour and reproduction can be induced by confining the animals in blacked-out pens for appropriate periods and replicating autumnal daylengths, and it is conceivable that this technique may eventually be utilized commercially as a means of increasing lamb yields in a similar manner to the light-induced increase in egg production exercised by the poultry industry.

In the southern hemisphere sexual activity of such animals is likewise correlated to the months of decreasing daylengths (March to August), and when they become transferred to the northern hemisphere they soon besome resynchronized and respond to the northern autumn, thus reversing their breeding pattern.

There is an inverse relationship between latitude and the duration of breeding. Thus, in sheep indigenous to countries in higher latitudes, the sexual season is much shorter and more marked than that of a breed which

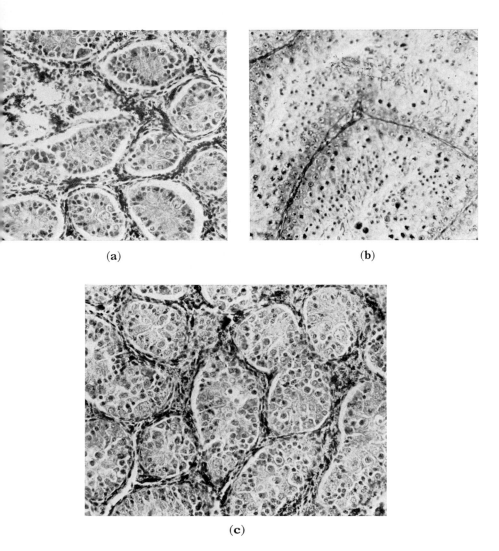

(a)

(b)

(c)

Plate 3 (a) Testis of stock dove in December. (b) Testis of stock dove caught in December and kept under an artificially simulated January photoperiod. (c) Testis of wood-pigeon kept under the same photoperiod conditions as (b). The latter failed to respond. (LOFTS, B. and MURTON, R. K. (1967). *J. Zool. Lond.*, **151**, 17.)

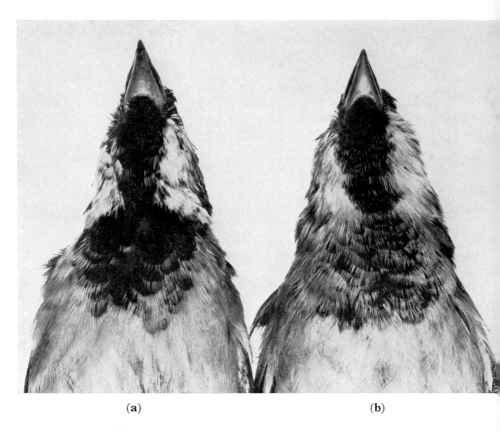

(a)   (b)

Plate 4 The effects of photoperiod on moult in the house sparrow. Specimen (**a**) was caught during the breeding period and held on 16-hour photoperiods for three months. Specimen (**b**) was kept under natural light conditions for the same period. The white-tipped new feathers are clearly seen. (LOFTS, B. and MURTON, R. K. (1968). *J. Zool. Lond.*, **155**, 327.)

has originated from progenitors indigenous to more tropical or equatorial parts of the world. An example of this is shown in the comparison of the reproductive cycles of the Suffolk and Merino breeds plotted in Fig. 3–1. The ancestors of the modern Merino sheep evolved in a relatively benign environment where there were no strong seasonally recurring selective factors, and lambs born at any time of the year had a chance of survival. In this breed therefore, the period of anoestrus is extremely brief, and the ewes can undergo up to 20 consecutive oestrous cycles in a year and, to all intents and purposes, remain capable of breeding throughout the whole of this period. The Suffolk ewes, on the other hand, are much more seasonal, and, in its more northerly and harsher environment, only the lambs born in the restricted period of the spring have a good chance of survival. In this breed, the period of anoestrus is very much greater, and the number of potential consecutive oestrous cycles very much fewer.

3.3 Photoperiod and delayed implantation

Light is known to have an influence on the phenomenon of delayed implantation. In many species of mammals after mating has occurred and the egg fertilized, there occurs a delay before the blastocysts become implanted in the uterus. The delay can vary from a few weeks to several months depending on the species. The phenomenon occurs in seals, bears, badgers, roe deer and some marsupials, but is most common in members of the Mustelidae, the mink and its relatives. It is an important means of prolonging pregnancy over the less propitious months of the year so that the young are born in spring and have a greater chance of survival. In roe deer, after mating in July or August, the doe enters a long period of delayed implantation which ends in December. The blastocyst then becomes implanted and starts its embryonic development, culminating in the birth of the kids in May. In the absence of a period of delayed implantation, birth would take place in December when winter conditions would probably prove fatal to the young roe deer.

In the marten, sable and mink it has been demonstrated experimentally that long days decrease the duration of delayed implantation. Thus, extra light given in winter to the long-tailed weasel, reduces gestation time by as much as three months. This knowledge has great practical importance in the commercial rearing of mink and sable, and artificial manipulation of the photoperiods can increase the number of pregnancies and yield of offspring.

3.4 Moult and antler cycles

The moulting of hair and the subsequent development of a new coat is sometimes influenced by the natural light changes. Many mammals shed their hair at the end of the summer concomitant with the regrowth of a much thicker winter coat. This is very evident in some species where the winter pelt is often a different colour from the summer one, but is also true of many species where a change of colour does not occur. The mechanism is co-ordinated by the environmental photoperiod: decreasing daylengths stimulate winter pelt development and increasing daylengths synchronize the spring moult and the assumption of the summer fur. Silver foxes are sometimes experimentally subjected to a period of very long photoperiods followed by a phase of decreasing photoperiods to induce an accelerated moult cycle and produce an earlier development of the prime winter pelts. Cattle which are exposed to 13-hour photoperiods in spring and autumn react by shedding their hair, but a similar treatment during the intermediate summer months will induce a rapid growth of hair without any shedding.

In many deer, antler growth is a cyclic event which is regulated by an interplay of light-controlled mechanisms mediated via the pituitary gland and the hormonal secretions of the testes. Antler growth starts in roebuck deer under the lengthening days of spring before the testes have reached full breeding condition. When the latter becomes fully developed and androgen secretion has reached a high level, the sex hormones terminate the further growth of the antlers, and cause the shedding of the 'velvet' covering so that they become essentially dead organs. 'Velvet' shedding occurs in March. After the rutting season is over and the secretion of gonadal hormones diminishes with the post-nuptial regression of the testes, the antlers are shed. Regrowth will start with the return of the lengthening days in the succeeding spring.

Photoperiodism in Lower Vertebrates   4

In poikilothermic vertebrates the temperature of the body changes when the environmental temperature varies, and since the metabolic activities slow down proportionally to the ambient temperature, such animals are much more susceptible to seasonal thermal fluctuations than homoiothermic species. It is not surprising, therefore, that *thermoperiodism* (i.e. the regulation of reproductive and other cycles by seasonal changes in environmental temperature) is sometimes dominant to photoperiodism. Examples of photoperiodic mechanisms are known, however, and, no doubt, as greater research effort becomes directed towards these lower vertebrates, many more will come to light. In short, the phenomenon is probably considerably more widespread than our present knowledge indicates.

4.1 Photoperiodism in fishes

Fishes, through the course of their evolution, have adapted themselves to a wide range of external stimuli that co-ordinate their seasonal breeding activities and migrations. In some, such as those species inhabiting the almost total darkness of the ocean depths, or living in subterranean waters like the blind cave fish, *Anoptichthys jordani*, light has no effect, but species living in bright light near the surface of the water can be strongly influenced by photoperiods. It is well known among keepers of home aquaria, for example, that the tropical guppy and swordtail, although capable of breeding all the year round, show a heightening of reproductive activity with the increasing daylengths of the spring.

It can be shown experimentally that light can affect the state of the reproductive organs in many species. Thus, goldfish kept in the dark for a long time show a degeneration of the gonads. Conversely, fishes belonging to several different families can be induced to display and show an acceleration of gametogenesis in response to artificially increased photoperiods, a technique that is widely employed in commercial trout fisheries where young fishes are induced to ripen many months in advance of normal by subjecting them to protracted photoperiods. The minnow, *Phoxinus laevis*, and stickleback, *Gasterosteus aculeatus*, are two common British species which can respond to such light treatment and provide useful experimental animals for simple laboratory experiments.

Although there is ample laboratory evidence that a gonadotrophic response to artificially protracted photoperiods is shown by many species of fish, it does not necessarily imply that environmental photoperiods are always the important factor influencing the natural reproductive cycles. In the few species that have been studied in detail it appears that light and heat impose a dual control. This is well demonstrated by the reproductive cycle of the common European minnow, *Phoxinus phoxinus*. This fish breeds in May and June, and immediately following, there is a re-establishment of gametogenetic activity in the gonads. The proliferation of new germ cells continues throughout the autumn then comes to a halt during the winter months. With the arrival of spring there is a recrudescence of gonadal activity and a rapid completion of maturation in both males and females. For either sex to attain full maturity requires a combination of long days with high temperatures. Thus, the increasing daylength and water temperature in the spring accelerates the fishes into full breeding condition. Without this combination, maturation of eggs in the ovary is arrested and spermatogenesis in males is held back at the spermatocyte stage. With short daylengths reproductive development is inhibited even if temperatures remain high and, similarly, a long daylength if combined with a low temperature, although sufficient to accelerate a winter minnow into full breeding condition, retards egg development which continues very slowly up to the early stages of yolk accumulation. Inhibition of egg development by high temperature in the absence of long days is widespread in fishes and in addition to *Phoxinus*, is also found in *Notropis*, *Rhodeus* and *Apeltes*. In all these forms, the inhibition by high temperature can be over ridden by long photoperiods.

Reproductive cycles which, like that of the minnow, are at least partly photoperiodically regulated, also occur in other cyprinid fish such as the bridled shiner, *Notropis bifrenatus*, the bitterling, *Rhodeus amarus* and the centrarchid banded sunfish, *Enneacanthus obesus*. In all these species, as in the minnow and stickleback, an inactive winter vegetative period, is followed by the photosensitive pre-spawning phase during which the gonads reach full sexual maturity, and the fishes develop nuptial colours and display sexual behaviour culminating in fertilization of the eggs. By transposing the Argentine viviparous fish, *Jenynsia lineata*, into the northern hemisphere it is possible to induce two breeding seasons within a single year, one during the southern hemisphere spring and summer (October to February) and one in response to the northern spring and summer (May to July).

In the bridled shiner, *Notropis bifrenatus*, the first visible sign of sexual development is the appearance of the nuptial colouration which usually first becomes apparent during the third week of April. Spawning takes place in June, after which the nuptial colours start to fade and have usually disappeared by the first week in August. There are roughly 46 days between the first outward signs of sexuality and the onset of spawning. This pre-spawning phase is advanced by long photoperiods given during the winter. For example, bridled shiners exposed to long days from January 1st, spawn in mid-February, some 44 days later. The experiments of R. W. HARRINGTON have shown that when long photoperiods are begun in mid-September, on the other hand, an interval of 97 days occurs before spawning starts. This indicates that the internal mechanisms governing functional maturity have a refractory period after the natural spawning phase which defers the response to long photoperiods and consequently, the subsequent spawning develops some 52 days late. Gonad histology confirms these results and show that no gametogenetic activity can be stimulated in the 'spent' gonads, until the fish has emerged from its refractory period. A post-spawning refractory period is thus an integral part of the reproductive cycle in several species of cyprinid fishes, as it is in many birds.

Unlike the fishes so far mentioned, some species like the brook trout, *Salvelinus fontinalis*, spawn in the autumn and respond to decreasing daylengths. In such a fish sexual maturity can be accelerated, or conversely, retarded, by experimentally reducing or maintaining the fish under long summer daylengths respectively. By shortening the summer daylengths in this way the sexual season can be advanced by a month. If the artificial shortening of the photoperiods is preceded by an abrupt period of increasing daylengths (i.e. in effect a contracted version of the annual light cycle), this experimental manipulation can advance maturity by up to four months, and is sometimes used in the commercial trout hatcheries.

An interesting example of the effects of light on fish, which also has a commercial application, is the phenomenon of 'runting' in species of the West African mouth breeding fish, *Tilapia*. In many parts of Africa and in other underdeveloped parts of the world where there is a deficiency of protein, culturing of *Tilapia* has been introduced for food production as an alternative source of protein. Unfortunately, it was discovered in the earlier attempts that transference of the fishes from their natural habitat into the shallow artificial canals and ponds for culturing exposed them to increased illumination which accelerated the development of sexual maturity and caused them to breed at a much earlier age than normal. The outcome was that the ponds soon became populated with young fish that had not yet become fully grown ('runts') and which were uneconomical and worthless as food. In Madras, for example, it was found that *T. mossambica* bred when they had only grown to a length of about two inches when kept in water less than a foot deep. The difficulty was overcome by increasing the depth of the culture ponds so that the deeper water reduced the intensity of illumination. In a depth of five feet fishes grew to a length of twelve inches before breeding.

In its natural environment of West Africa, the reproductive activity of *Tilapia* decreases during the period of heavy rains in June and July, and again in the October-November 'small rains'. The rains produce deeper and more turbid water which has the effect of reducing the illumination of the fishes' habitat and this, in turn, depresses the breeding activity. Also, due to flooding of the lowland coastal areas, the fishes become spread into waters with dense vegetation, which again diminishes the light.

Photoperiodic regulation is not exclusively confined to the control of the reproductive rhythms, but can also be influential in other physiological processes. Many fishes at the time of migration show changes in the activity of the pituitary, thyroid and interrenal glands, and changes in metabolism and general activity. The three-spined stickleback undertakes a pre-spawning migration from the sea to fresh water during the spring, and a post-spawning migration back to salt water during the autumn. By experimentally exposing the fish to a range of salinities throughout the year, B. BAGGERMAN has demonstrated that at the time the fishes are scheduled to start their spring migration they show a seasonal change in preference from salt to fresh water, but in the autumn when they start the downstream journey, there is a return of preference for more saline waters. The underlying physiological mechanisms are not known in detail but are thought to involve the pituitary and thyroid glands which are in turn affected by photoperiods. Thus, the long days of spring induce a fresh water preference, whereas the declining days of autumn reimposes a preference for more saline waters. Experimentally, cyclical changes in salinity preference from salt to fresh water can be induced in these fishes by maintaining them under long photoperiods and a constant high temperature, whereas exposure to short photoperiods under similar temperature conditions, induces the animals to maintain their initial salt water preference. Thus, the daily photoperiods control the time at which changes in salinity preference take place. It is unlikely, of course, that these seasonal changes in salinity preference are responsible for such a complicated behavioural pattern as migration, but it indicates the time of the seasonal induction of migration-disposition.

The Pacific salmon, *Oncorhynchus*, is another species that migrates from the sea into fresh water to spawn. In the Coho (*O. kisutch*) and sockeye (*O. nerka*) the young fry hatch out and remain in fresh water for a year before migrating seawards. As in the stickleback, this departure is presumably induced by the seasonal assumption of a migration-disposition, indicated by a change in salinity preference, and timed by a photoperiodic mechanism. But whereas in the stickleback a long daylength induces a freshwater preference, in juvenile fresh water salmon it causes a salt water preference just before their seaward migration. In fishes kept artificially on short daylengths this change in preference from fresh to salt water is inhibited.

4.2 Photoperiodism in amphibians

The majority of amphibians periodically return to water for their breeding activities. Consequently, the timing of their annual reproductive cycles is, to a large extent, dependent upon conditions such as rainfall, availability of ponds or streams, or such similar factors which regulate the water supply. As poikilotherms, they are also influenced by external temperatures. It has been claimed that light can have an effect on the release of mature spermatozoa from the testicular seminiferous tubules in some species, such as the African clawed toad, *Xenopus*, but generally evidence for photoperiodic controlling mechanisms in this group of vertebrate animals is sparse.

Amphibian species appear to be little affected by seasonal variations in daylength. Some early claims that frogs, newts and salamanders respond to photoperiodic stimulation, and that a prolonged exposure to a summer photoperiod in winter induces an unseasonal spermatogenetic recrudescence, can be ascribed to the effect of a high room temperature on animals removed from their natural winter habitat and placed in a laboratory. Even in complete darkness an elevation of the ambient temperature will stimulate gametogenesis in such animals.

When the common frog, *Rana temporaria*, is kept in the dark for long periods, the gonads can still develop into full breeding condition. Furthermore, exposure to a steadily increasing, or excessively long, photoperiod has no accelerating effect. Spermatogenesis has been shown to be completely normal in several species which have been kept in complete darkness from their embryonic stages. The newt, *Triturus alpestris*, is an example.

In the amphibians then, our present state of knowledge suggests that photoperiodic mechanisms regulating annual reproductive cycles, appear to be absent, and temperature and rainfall are the main environmental synchronizers.

4.3 Photoperiodism in reptiles

The reptiles are fully adapted terrestrial vertebrates and, unlike the majority of amphibians which are still dependent upon an aquatic habitat during at least part of their life cycle, generally live and reproduce under the conditions imposed by a land environment. But whereas the two other groups of truly terrestrial vertebrates, the birds and mammals, have enjoyed the great liberation provided by 'warm-bloodedness', reptiles are poikilothermic and like the other poikilotherms already considered, the effects of daylength are often inextricably intertwined with those of temperature. Photoexperimentation on reptilian species has been lacking, but although our information is meagre, there is some evidence that the reproductive processes, in some species at least, can be influenced by light.

Fishes and aquatic amphibians occupy an environment in which the rate and extent of temperature fluctuations are minimized by water. This is very different from the conditions of many reptilian species that live in semi-arid or desert areas where the seasonal and daily temperature changes can be extreme. In evaluating the role of daylength in the synchronization of the breeding season, one cannot rely exclusively on laboratory data, but knowledge of a great deal of the natural history and behaviour of the animal involved is also necessary.

Reptiles maintain their body temperature at a characteristic level and this can vary from species to species. The desert iguana, *Dipsosaurus*, for example, is active at body temperatures in excess of 43°C, a level higher than that of most birds and mammals. Such a temperature, however, would prove lethal to many other reptilian forms. In fact, reptiles can withstand very little deviation from their individual body temperature thresholds, far less than most poikilothermic animals. This lack of tolerance to temperature fluctuations, and the absence of an effective physiological thermoregulatory mechanism, makes them very susceptible to environmental temperature changes and they have evolved patterns of behavioural activities which enable them to minimize the effects of such variations. These, in turn, will determine the pattern of their exposure to light. Thus, desert lizards in general are normally active during the morning in late spring and summer, but they retreat underground when the mid-day sun elevates the environmental temperatures above a critical level. They reappear above ground in the late afternoon when it becomes somewhat cooler. Some species, like the crested lizard, *Dipsosaurus dorsalis*, may remain sheltered from the sun for nearly the whole day in the height of summer. With such behaviour the amount of light the animals receive will be restricted, and their exposure to it will be determined by their thermoregulatory behaviour and not by sunrise and sunset. Nevertheless, photoperiodicity can still be a primary synchronizing factor, and environmental temperatures generally increase with longer days and hours of sunshine.

One species of desert lizard which has been studied in some detail is the fringe-toed lizard, *Uma notata*, which lives in the deserts of Southern California and Arizona. Its habitat is one of loose sand and when disturbed it rapidly burrows until it is completely concealed. Its toes, as its name implies, are fringed and this aids it in moving over the loose sand. During the cooler winter months the animals hibernate and remain underground for the greater part of the day, only emerging for a short period of activity during the warmer mid-day period. With the advent of the longer, sunnier days of early spring they emerge for progressively longer periods of activity and thereby experience an increasing number of daylight hours from January to the end of April. Hours of available sunshine range from approximately 10·25 hours in January to 13·5 hours in April. During this period the increasing photoperiod stimulates the development of the gonads, recrudescence of gametogenesis beginning in March and culminating in breeding and egg-laying in April. Breeding behaviour then declines, in spite of the progressively lengthening photoperiod because the high environmental temperature during the early afternoon from May to September force the lizards to retreat underground. Consequently the length of the actual photoperiod to which the animal is exposed is reduced although environmental daylengths are increasing (Fig. 4–1). During July they only receive approximately 9 hours of light per day.

The photoperiodic responses of *Uma* have also been studied experimentally, and an unseasonal development into breeding condition in both sexes can be induced by photostimulating winter lizards with a summer daylength. A similar gonadotrophic response can similarly be induced in a variety of reptilian species including turtles, chameleons, lizards and snakes. However, where such photoexperimentation has shown a photoperiodic mechanism, generally knowledge of the natural behavioural activities has been largely ignored and therefore such results must be treated with caution when interpreting them as proof of a cycle regulated by natural seasonal photoperiods.

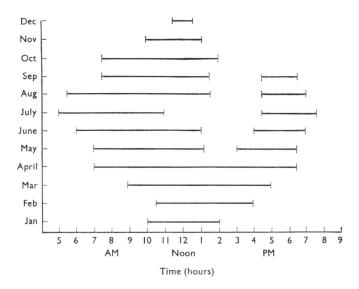

Fig. 4–1 The duration of time spent above ground by the fringe-toed lizard at different times of the year. (MAYHEW, W. W. (1964). *Herpetologica*, **20**, 95.)

Reptiles that inhabit humid tropical areas do not have the thermoregulatory problems of those species inhabiting arid areas and photoperiodism has a much more direct regulatory role in the seasonal events. In the tropical rainforests seasonal changes in rainfall and temperature are often remarkably small and species like the *Emoia* lizards breed during the longest days of the year in the New Hebrides and show their minimal reproductive activity during the period of shortest days.

Daylength not only influences the reproductive cycles, but, in at least one species of desert-dwelling lizard, *Dipsosaurus dorsalis*, can also affect food consumption, appetite and growth. Thus, if newly hatched animals are exposed to prolonged illumination, they grow much more rapidly.

Furthermore, winter animals gain considerably more weight than summer animals exposed to the same conditions, so that during the hibernation period the metabolic processes either slow down or become more efficient, and body food reserves build up. The feeding habits also change with the photoperiod; appetite is considerably greater during the spring and summer months than in the autumn and winter. Thus the increase in body weight of winter animals occurs when the appetite is reduced and the food intake consequently decreased. This feeding response is, apparently, typical of lizards generally and probably represents an adaptation of some physiological processes in preparation for hibernation.

Photoperiodism in Invertebrates   5

Within the last two decades considerable interest has been shown in invertebrate photoperiodism, and sufficient data have now accumulated to show that the phenomenon is very widespread, particularly in terrestrial arthropods. Evidence of seasonal light-controlled cycles in marine invertebrates is inconclusive, which is perhaps not surprising since the oceans constitute a highly constant environment, where seasonal temperature changes and tidal fluctuations provide consistent synchronization. In terrestrial species, however, particularly insects and mites, the seasonal and diurnal photoperiods have a significant regulatory role, and these animals have exploited environmental light fluctuations in the evolutionary adaptation of their physiological and behavioural mechanisms to just as significant a degree as have the vertebrate species. As in the latter, insect reproductive cycles are adapted to produce progeny during the clement months of the year, and in species occupying areas where the summer season is short, and the winter climate relatively rigorous, only one generation may be produced annually. In species inhabiting environments marked by more favourable climatic conditions, however, more than one generation may be propagated.

To avoid the rigours of the winter months species enter a state of hibernation during which the vital activities and many metabolic processes either stop or become retarded to a very low level. This state is known as *diapause*, and is one of the seasonal events which is regulated by a photoperiodic mechanism. It is a phenomenon that is extremely widespread in insects and probably forms an integral part of the life history of most species inhabiting the temperate and colder parts of the world. It is not a rhythmic event, however, and generally occurs but once in a lifetime. Another light-regulated mechanism displayed by some forms is the phenomenon of polymorphism, where the one species may display two seasonal forms or, as is generally found in aphids, the alternation of several morphologically dissimilar generations within one annual cycle.

5.1 Timing of diapause

The photoperiodic induction of diapause has been investigated in over a hundred species. It is an adaptive mechanism of great importance to the animal which, in some ways, might be considered analogous to the refractory period in vertebrate reproduction since it prevents the possibility of reproduction at times when periods of extreme summer heat, or dryness, or low winter temperatures might produce a high mortality of offspring. The specific point in the life cycle at which it occurs varies from species to species, and may be undergone at any phase of development. Thus, it may involve the egg, nymph, larva, pupa or adult stages. Within any particular species however, the particular growth stage at which it occurs is a genetically determined factor, and generally one stage only is involved. It becomes manifest well before the onset of unfavourable physical conditions.

Moisture, diet and temperature can all influence diapause, but these are generally of secondary importance, and in most of the species so far investigated, environmental photoperiods play the decisive role. In some, diapause is an obligatory part of the life history of every individual animal in each generation, but, more commonly, most species can produce individuals in which a diapausal state may or may not develop, depending on the environmental conditions prevailing during the time of photoperiodic induction. The latter condition is known as a *facultative diapause*, whereas the former is called an *obligatory diapause*.

The point in the life history at which the photoperiod is effective as a diapausal synchronizer generally preceeds the period during which the diapausal state is actually manifested. For example, in those species where survival during an adverse season is brought about by means of an embryonic diapause in the egg stage, the induction of the diapause is determined by the light conditions experienced by the parent generation. The silkworm, *Bombyx mori*, is an example of such a species, and in this animal whether or not a diapause will occur in the eggs is determined by the photoperiods experienced by the female parent during its late embryonic and early larval life. In species which show an adult diapause, its induction is most frequently determined by the photoperiods experienced by the insect during its larval stages, and, in many cases, it appears primarily as a suppression of reproduction.

In the experimental investigation of the photoperiodic induction of diapause the effect is assessed by determining, in any particular species, the percentage of the population which is induced into a diapause whilst being reared under controlled conditions of diet, temperature and photoperiod. By repeating the experiment under different light regimes, it is possible to determine the effect of different photoperiods on the percentage incidence of diapause, which can then be plotted as an induction response curve.

5.2   Long-day and short-day forms

Of the arthropods which have so far been investigated by means of the above techniques, the effectiveness of a range of photoperiods to induce a state of diapause in any particular insect or mite has been shown to vary considerably from species to species. On the basis of their diapause induction response curve they can generally be classified into two basic types, which are illustrated in Fig. 5–1.

Type A is the type of response which has been found in a large number of species, and is known as a *long-day response*. In these forms a long daily photoperiod allows uninterrupted growth and development, whereas a period of short days will initiate a state of diapause. The actual length of the critical photoperiod is, of course, a species-specific factor which has been determined by natural selection, and is usually quite sharply defined. In the colorado beetle, *Leptinotarsa decemlineata*, where the diapause occurs in the adult form, the critical photoperiod is 14 hours, but in the pitcher-plant midge, *Metriocnemus knabi*, the critical photoperiod is about 12 hours, and any daylength below this threshold prevents pupation and instead

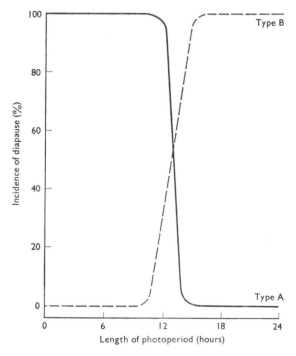

Fig. 5–1   (A) Long-day and (B) short-day types of diapause response to the length of the photoperiod.

induces the development of a diapause larva. In most species so far studied, the critical daylength range generally lies somewhere between 12 and 16 hours. If these 'long-day' types are maintained artificially under long photoperiods they never enter a diapausal state. The survival value of such a mechanism can be seen by examining the life cycles of the red spider mite, *Metatetranychus ulmi*, which shows an embryonic diapause that is determined by the photoperiods experienced by the female parent. Being a long-day type the diapause is induced when the female in its deutonymphal stage experiences short daylengths. In the spring the mite population reaches maturity when the natural photoperiods are long so that females lay eggs which are non-diapausal and develop immediately to produce a further generation. This is repeated, but as the season progresses later generations reach their deutonymphal and early adult stages during the declining day lengths of late summer. The females then lay eggs which enter into diapause, and only recommence their development with the advent of the long sunny days of the following spring. Thus this photoperiodic adaptation synchronizes the biology of the mite with that of the

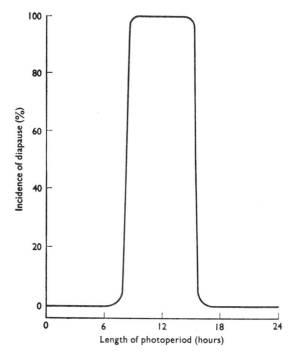

Fig. 5-2   Diapause response in a short-day/long-day type insect.

host plant and the annual climatic conditions, and several generations may be produced when environmental conditions are most favourable.

Type B is the induction response curve of a *short-day* forms and is the reverse of the long-day pattern. It is far less common than the latter and is found, among others, in the silkworm *Bombyx*. Insects with this type of photoperiodic response, show uninterrupted development under short photoperiods, whereas an exposure to long photoperiods will cause an induction of diapause. *Bombyx*, like the red spider mite, has its diapausal state as an egg, and the photoperiodic experience of the female adult determines whether an embryonic diapause is induced. Under normal circumstances the over-winter eggs hatch in early spring when photoperiods are still relatively short, and larval development under these light conditions culminates with the emergence of the moth in early summer. These adults lay eggs during the period of long days, so that the subsequent late embryonic and larval development of this generation takes place under long photoperiods and the moth, which eventually emerges in the autumn, deposits eggs that pass the winter in diapause. The short-day forms, therefore, differ from the long-day types in that they are limited to two generations per year, because the long summer photoperiods cause a diapausal induction in the next generation. Furthermore, in this type of response, under conditions of continuous illumination or continuous long photoperiods, diapause is always effected.

In addition to the two main types of responses outlined above, a number of insect species show a variation on these two patterns. An example of this is seen in the European corn borer, *Ostrinia nubilalis*, whose induction response curve is shown in Fig. 5–2. In this form, development occurs without interruption under conditions of both short and long photoperiods, and diapause is only induced when animals are exposed to a relatively narrow range of photoperiods of between about 8 to 16 hours duration. Daylengths shorter than 8 hours or longer than 16 hours will not induce diapause. This is sometimes classified as a short-day/long-day response and, in addition to *Ostrinia*, is also known to occur in the oriental fruit moth, *Grapholitha molesta* and cabbage white butterfly, *Pieris brassicae*. It is difficult to see any ecological value in such a response, since under field conditions 8-hour photoperiods or less are only experienced in the middle of winter, and at a time when the insect would already be in a diapausal state.

In a few lepidopteran species an interesting variant occurs which is the converse of the short-day/long-day type. In these, diapausal induction is produced by both long and short photoperiods so that the induction curve is the reverse of that shown in Fig. 5–2. The peach-fruit moth, *Carposina niponensis*, is an example of such a form.

Although the photoperiod is known to be the primary inducer of diapause in many insects, it has generally no influence in the termination of this state in most of the species so far studied. However, in a few forms, there is evidence that light can accelerate the termination of diapause. Thus, in some mosquitoes like *Anopheles barberi* and *Aedes triseriatus*, exposure to long photoperiods will effectively end the larval diapause. The adult diapause of the colorado beetle can also be terminated by exposing the insects for several days to long photoperiods. In the short-day type *Limnephilus* species, on the other hand, a premature termination of diapause is provoked by keeping the insects under a short-photoperiod regime.

5.3 Temperature modification of the photoperiodic response

The influence of temperature on the induction of diapause is variable. Generally, in insects which display a long-day response (Type A), high temperatures tend to shift the induction response curve to the left, whereas a decline in the ambient temperature shifts it to the right, i.e. the length of the photoperiod which induces diapause varies inversely with the ambient temperature. Thus a species such as the moth, *Acronycta*, can take advantage of an unseasonal spell of warm weather in late summer, and produce an additional generation, because the critical photoperiod that induces a diapause will be shorter as a result of the modifying role of temperature and, therefore, encountered later in the season. Conversely, those arthropods that are of the short-day type will show the reverse effect, and relatively high ambient temperatures tend to promote diapause induction. In some species there appears to be a critical temperature preference for the photoperiod to be effective. Thus, the diapause-inducing photoperiod of *Metriocnemus* is most effective when the environmental temperatures are between 23°C and 25°C, and if they are above or below this range they retard the rate of response.

Temperature can therefore cause a marked modification of the effect of photoperiods and, in some species, the photoperiodic induction of diapause may be completely abolished by high environmental temperatures. Such is the case with the cabbage worm where insects reared under experimental conditions of continuous darkness do not enter a pupal diapause when environmental temperatures are in excess of 20°C. Under more natural light conditions of a 12-hour daily photoperiod it requires a temperature level of above 28°C to completely avert the diapausal response. In contrast, the moth *Dendrolimus*, the silkworm *Autheraea*, and the mosquito *Culex pipiens*, are species in which the photoresponses are relatively independent of the environmental temperatures. Most insect species, however, need to be exposed to a period of low temperatures before they can emerge from a diapausal state and resume their active growth and development.

In some species, the interaction of photoperiod and temperature influence the intensity of the diapause resulting in a transient summertime diapause (aestival diapause) under conditions of long daylengths and high temperature, and a more intense long-enduring winter diapause, induced by short days and low temperatures. This is the case with the cabbage-moth, *Mamestra brassicae*, where the photoperiods and temperatures experienced by the growing caterpillars will determine the intensity of the diapause which is effected during pupation.

5.4 Photoperiodic adaptation in geographical races

The diapausal photo-threshold which needs to be exceeded to allow continued development and prevent diapausal induction in long-day types, can vary with latitude in species that have a wide geographical distribution, the tendency being for the photo-threshold to be longer in the higher latitudes. In England for example, adult red mites enter their winter diapause in response to a natural daylength which is two hours shorter than the critical photoperiod which causes a Leningrad population of the same species to enter the same state 8° further north. In Russia, A. S. DANILYEVSKY has carried out a detailed investigation of this effect on the photoperiodic responses of *Acronycta*. His results are summarized in Table 3, and they clearly demonstrate how populations of this moth col-

Table 3 Relationship between latitude and the photoperiodic induction of diapause in populations of the moth *Acronycta*.

Location	Latitude	Critical photoperiod (hours)
Leningrad	60	20
Vitebsk	55	18
Byelgorod	51	17
Sukhumi	43	14·5

lected from different localities, show a considerable variation in their photoperiodic induction of diapause under experimental conditions with constant temperature. At a latitude of 60°N the moth responds to a critical photoperiod of nearly 20 hours (which may be a short day at this latitude) so that any daylength falling below this threshold will induce a diapause, but at a latitude of 43°N, daylengths have to reduce to 14·5 hours before the same species enters diapause. Thus, the more northerly population that occupies a latitude where daylengths are very much longer than at 43°N enters a diapausal state at a photoperiod which prevents the induction of this state in the more southerly moths which live under conditions of much shorter environmental daylengths. The overall effect therefore, is that the further south one moves, the shorter the photoperiod becomes that is necessary for the continued development and growth, the net result being that pupae are released from diapause much earlier so that three generations can develop under the long-day conditions that persist for more than four months in these latitudes, whereas only a single generation may be produced in Leningrad. A further difference noted by Danilyevsky in these northern and southern populations, is that, whereas rearing the southern populations at an experimental temperature of 30°C can counteract the diapausal induction effect of a short day, it cannot avert the short-day induction of diapause in the Leningrad population.

    We have already mentioned that *Mamestra* is a species in which a summer aestival diapause may sometimes occur in addition to the normal
4 + A.P.

winter diapause, and the Japanese races of this species show a clear adaptation of this phenomenon to the environment at different degrees of latitude. S. MASAKI has shown that strains from northern Japan show the usual photoperiodic response, a short day inducing an intense diapause and a long daylength preventing it. In these relatively cool latitudes (45°N) the spring emergence of adults from the over-wintering pupae does not occur until July. There are usually two generations produced, and all of the larvae of the second generation develop under declining daylength conditions so that the pupae are committed to a winter diapause. An aestival diapause rarely occurs. But moths taken from the populations in the south show an aestival diapause which shields them from the high summer temperatures and more arid conditions that seasonally prevail at these latitudes (35°N). Thus, spring adults emerge in early May, and the first generation of progeny pupate in June and enter a transient diapause that suspends further development until the cooler months in late summer. The adult moths of this generation emerge in late August and early September to propagate a second generation which develop under declining daylengths and temperature, and pupate into an overwinter diapausal state. If such an aestivating pattern should have occurred under the more abbreviated summer season of the northern population, it would mean that the second generation would not have time to develop before the onset of winter conditions, and would result in a high mortality rate. Selection pressure has therefore, eliminated this mechanism from the life history of these moths. Between these two extremes of latitude, Masaki has been able to demonstrate that populations of cabbage moths reared under experimental conditions of 16 hours photoperiods and 25°C, show an increasing percentage incidence of aestivation in the first generation pupae, as the collection of moths under investigation are made from more and more southerly habitats.

5.5 Photoperiodic control of body form

A number of insect species exhibit striking seasonal changes in their morphological appearance, which in many instances in the past, has led to much confusion in their taxonomic classification. These changes are often regulated by a photoperiodically controlled mechanism. A classical example of this phenomenon is seen in the European nymphalid butterfly *Araschnia*, which was originally described as two distinct species, *A. levana* and *A. prorsa*. The latter form, in fact, is the adult stage of the summer generation, is black with white spots, and is quite distinct from the red, black-spotted levana form which develops from the over-winter diapausal eggs. The differentiation of these characteristic wing patterns is connected with the photoperiodic experience of the larval growth stage, and if larvae are exposed to long photoperiods the pupae always give rise to the *prorsa* form, whereas rearing caterpillars under short photoperiods leads to the induction of a diapause pupa and the subsequent emergence of the *lavana* form of butterfly. Several lepidopteran insects show a similar seasonal dimorphism and as in the example already considered, the alternation can usually be associated with a diapausal and non-diapausal generation. There may be some selective value in this dimorphism in that the two different colourations adapt each respective generation, in some species at least, to blend better with its environmental background at that particular time of year. Whereas among lepidopteran species such dimorphism is usually associated with morphological variation in wing pigmentation, in some homopteran species it also results in the expression of different wing-lengths which can again be correlated with whether the adult has emerged from an over-winter diapause stage, or a non-diapause summer generation. In the pearsucker, *Psylla pyri*, for example, the winter generation is always larger and darker, and have longer wings than the summer generation. In the rice leafhopper, *Nephotettix cincticeps*, a small short-winged form develops from the diapausing nymphs in the spring, and is succeeded by a larger, long-winged summer generation.

Some of the most varied and complex forms of polymorphism are seen in the life cycles of aphids. In this group of insects polymorphism is very widespread and this phenomenon has been more extensively studied in these animals, than in any other group of insects. In these animals, photoperiodism controls the production of parthenogenetic and sexual forms. In a number of species the life cycle involves an alternation of host plants. The black bean aphid, *Aphis fabae*, for example, lives on broad bean plants during the summer but migrates to an alternative host, the spindle tree, for the winter, and this alternation of hosts is often associated with an alternation of the morphological appearance of the aphid. In its summer form the black bean aphid produces several generations of wingless, parthenogenetic females (virginoparae) which give rise asexually to further parthenogenetic forms. In late summer, however, the shortening daylengths influence the animals so that the production of wingless aphids finishes, and instead winged offspring (sexuparae) are produced which desert the summer host and migrate to the winter host. On the spindle tree the winged forms produce generations of oviparous wingless females (oviparae) and winged males. Sexual reproduction thus takes place in the autumn and results in the laying of diapause eggs on the winter host. In the following year the long days of spring and early summer cause the breaking of diapause and the hatching out of the eggs. The larvae develop into parthenogenetic wing-less females (fundatrix) which propagate several generations on the primary host plant, and eventually produce winged migrant females which return to the broad bean plants and re-establish the generations of par-thenogenetic, viviparous wingless forms. The fundatrix form is also known as the 'stem mother' and is always associated with the primary host plant. This life cycle is summarized diagrammatically in Fig. 5–3.

Some aphids do not have an alternation of hosts and show a different pattern of development from that outlined above. The vetch aphid *Megoura*

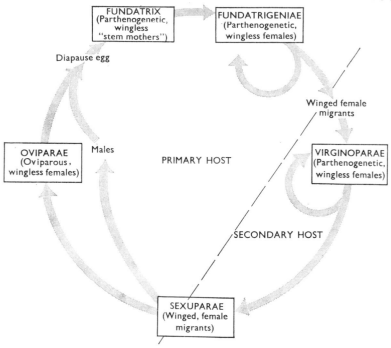

Fig. 5–3   Diagrammatic outline of the annual life cycle of the black bean aphid.

viciae is such a 'one host' species, and when the shortening daylengths cause the cessation of production of the summer parthenogenetic wingless virginoparae, it is not followed by the production of the migrant sexuparae, but immediately gives rise to the sexual male and egg-laying female oviparae. The factors which determine the production of virginopara and ovipara generations in this species has been studied in great detail by A. D. LEES. He discovered that maintaining cultures of these aphids at a constant temperature of around 15°C and a 16-hour photoperiod nearly always lead to the propagation of generations of virginoparae and never oviparous forms, but when first instar nymphs are transferred to a shorter daily photoperiod of 12 hours, although they develop into virginoparae, all the subsequent female progeny are exclusively oviparous. In this species, therefore, form determination appears to occur during the embryonic stages, and by further experimentation Lees was able to demonstrate that the critical daylength appeared to be around 14·5 hours, so that when virginoparae are reared from the first instar stage under photoperiods which are shorter than this critical daylength, they always propagate oviparous females, whereas virginoparae are always produced under day-length conditions in excess of this level.

Adaptive evolution of photoperiodically controlled seasonal processes has produced species-specific photo-thresholds which are the decisive factors in any photoperiodic reaction. Under normal night-day cycles the critical daylength has to be exceeded to initiate the relevant process, whether it be a recrudescence of gonadal activity, the induction of diapause in a 'short-day' insect, or an emergence from a diapausal state in those arthropods where the termination of diapause can be accelerated by light. Light periods shorter than the critical daylength are required for the termination of the refractory period in vertebrate animals or for the induction of diapause in 'long-day' insects. In the strongly photoperiodic species of northern latitudes, the threshold is sufficiently critical to make differences in photoperiods amounting to only a few minutes a day sufficient to initiate the process. Thus, stock doves respond to a January photoperiod and start their early spring gonadal recrudescence even though the difference in daylength from that of the non-stimulatory winter-solstice, is only a matter of 8 minutes, and similarly, in the pink bollworm where the critical photo-threshold for the induction of diapause is 13 hours, exposure to a photoperiod only 15 minutes longer will completely prevent the induction of a diapausal state. The precision of the timing of annual migrations, which can be predicted with an error of only a few days, further underlines the responsiveness of animals to such changes in daylength and hence, to the seasons. Such phenomena indicate the great accuracy with which many are able to 'measure' the natural photoperiod. Although a great many investigations have elucidated the photoperiodic effects in a wide range of both plant and animal species, relatively few have been designed to determine specifically what mechanisms are involved in time measurement. Nevertheless, a number of theories have been put forward.

6.1 Circadian rhythms and light sensitivity

Living organisms possess a physiological clock which they make use of for accurate time measurement. The basis of such a clock is an endogenous diurnal rhythm, and both plants and animals are known to possess many such oscillating systems. The results of the activity of these systems are generally referred to as *circadian rhythms* since the frequency of the oscillation (i.e. the time from one peak to the next peak in the 24-hour cycle) is rarely precisely 24 hours, but is slightly more or less. They were first discovered in plants when the existence of an endogenous rhythm in diurnal leaf movements was indicated in 1729 by the experiments of DE

MAIRAN. Since that time, particularly within the last two decades, many circadian fluctuations in physiological processes and activity patterns have been described. In recent years, under the stimulus of aeronautical and space research, there has been an intensification of interest in the biological clock governing many of the physiological rhythms of our own bodies. These rhythms are sometimes disturbed and put out of phase when people are rapidly transferred by plane from east to west or vice versa, and the effects that space flight might have on these endogenous processes is of great topical interest.

The phasing of circadian rhythms is determined by external factors; usually the daily alternations of light and dark serve to synchronize or entrain these endogenous oscillators. Experimentally isolating the organism from the natural external synchronizer (or *Zeitgeber*) by keeping it under conditions of constant light or darkness, causes the rhythm to become 'free running', and the true phasing becomes apparent. This technique demonstrates that such rhythms are rarely of a strict 24-hour periodicity. A circadian rhythm which is of longer duration will reach its peak slightly later each day, and if shorter, will do so earlier each day. Under natural conditions, however, the clock is 'set' by the daily change from dark to light at dawn, or from light to dark at dusk. In an earlier publication in this series, J. D. CARTHY likened a circadian rhythm to a clock which may be running a little too fast or a little too slow; the environmental 'cue' which entrains the rhythm can then be likened to the setting of the hands of the clock. The rhythms are independent of temperature fluctuations within the normal range encountered by any particular organisms.

That endogenous daily rhythms are somehow involved in estimating the length of the photoperiod was first proposed by E. BÜNNING in 1936. His theoretical model envisaged a circadian rhythm of cellular function (such as the secretion of gonadotrophin hormones) consisting of two half-cycles each of approximately 12 hours duration, one of which was 'light-requiring' (photophil) while the other was 'dark-requiring' (the scotophil). The photoperiodic induction of a process requiring long days occurred only when the duration of the natural photoperiod extended into the scotophilic part of the cycle. In its simplest concept, the model was envisaged as an endogenous rhythm of light sensitivity whose peak, within the 24-hour cycle, was phase-locked by the dawn. A large body of experimental evidence now supports the Bünning hypothesis in both plants and invertebrate animals, but only recently has it been implicated in the photoperiodic mechanisms regulating the seasonal cycles of vertebrates.

When North American house finches, *Carpodacus mexicanus* are caught in a sexually regressed winter condition, and are kept under daily light cycles of 6-hour photoperiods (i.e. 6L : 18D), this amount of light is well below the normal critical threshold for this species, so the gonads remain small and spermatogenetically inactive. When, however, such birds are divided into a number of separate groups, all of which are given 6-hour light periods, but coupled in each group with a dark period of 6, 30, 42, 54, or 66 hours respectively (ahemeral cycles), a clear-cut difference in the development of the gonads is observable after a period of 5 weeks (Fig. 6–1). Birds that have been subjected to 48-hour (6L : 42D) and 72-hour (6L : 66D) cycles in no way differ from controls kept under the original 24-hour (6L : 18D) regime, but the 12-hour (6L : 6D), 36-hour (6L : 30D) and 60-hour (6L : 54D) cycle groups respond as though they have been subjected to long daily photoperiods, and develop gonads of full breeding dimensions and spermatogenetic activity, even though the photoperiod is

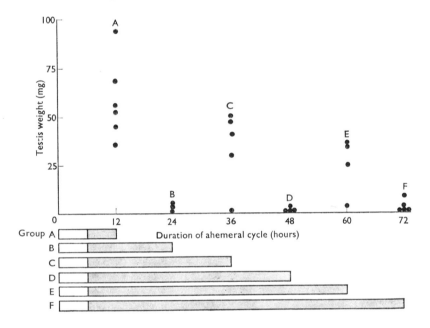

Fig. 6–1   Testis weights in house-finch kept under different ahemeral light cycles for five weeks. (HAMNER, W. M. (1963). *Science*, **142**, 1294.)

far below the normal photo-threshold. It is clear that such a response cannot support any hypothesis which stresses a requirement for a given duration of light or darkness in order to trigger a response. Although the length of the dark period differed in each case, it seems unlikely that this is itself important since birds subjected to 36-hour cycles show a gonadal response while birds exposed to less darkness (24-hour cycle) or more darkness (48-hour cycle) remain sexually immature. The only way the results can be explained is on the basis of an endogenous circadian rhythm of photosensitivity so that when light is experienced at the proper phase, gonadal recrudescence occurs. In short, it is not the quantity of light which is important, but where it falls relative to the light-sensitive rhythm. The phasing of this oscillation is 'set' by the change-over from dark to light at the start of the light period and this synchronizes the development of the light-sensitive phase to occur a certain number of hours afterwards.

Using similar experimental techniques to those outlined above, it is possible to demonstrate that circadian rhythms are involved in the photoperiodic gonadal responses of a variety of avian species. Figure 6–2 sum-

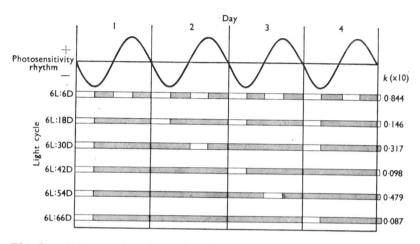

Fig. 6–2 The rate of testicular development (*k*) in groups of Japanese quail maintained under different ahemeral cycles. (LOFTS, B., FOLLETT, B. K. and MURTON, R. K. (1970). *Mem. Soc. Endocr.*, **18**, 545.)

marizes the response of the Japanese quale, *Coturnix c. coturnix*, kept under similar ahemeral light cycles. As in house finches, gonadal maturation under 12-, 36- and 60-hour cycles is significantly greater than under the other schedules. A theoretical curve of the light-sensitive rhythm is drawn at the top of the figure to illustrate that it is only the cycles in which the recurrent photoperiods coincide with the photosensitive period that maximum gonadal stimulation occurs.

6.2 The photoinducible phase

The period in the daily cycle during which the animal is light-sensitive so far as photoperiodic induction of diapause or gonadal recrudescence is concerned, is termed the *photoinducible phase*. It is only when the animal experiences light at this time each day, that a stimulation of gonadal development is induced. Light experienced at any other time in the 24-hour period, is ineffective. Although experimental ahemeral light cycles illustrate the presence of a circadian rhythm in the response mechanism, they have a drawback in that they do not give much information on the phasing and duration of the photoinducible phase. For this we depend upon another type of experimental technique in which the animals are kept under a nonstimulatory daily photoperiod (e.g. 6L:18D) and a second short light pulse is given as an interruption in the dark period. By varying the time at which the light pulse is given in different groups of experimental animals, the dark period or scotophase is effectively 'scanned' by the light pulse, and an analysis of the gonads of birds kept under such conditions for several weeks will indicate the position and duration of the photoinducible phase. The results of such an experiment carried out on *Coturnix* are shown in Fig. 6–3. Each experimental group was exposed to

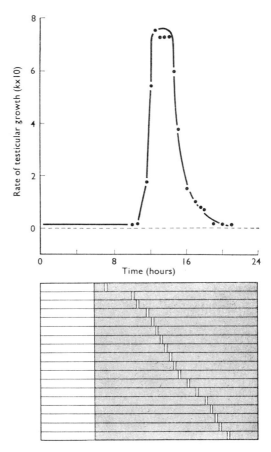

Fig. 6–3 Effect of 15 minute light-breaks on the rate of testicular growth in Japanese quail maintained under a 6-hour photoperiod (FOLLETT, B. K. and SHARP, P. J. (1969). *Nature, Lond.*, **223**, 968.)

4*

the same total daily ration of light of only 6·25 hours (6-hour photoperiod plus 15 minute light pulse), an amount which is below the photo-threshold of this bird if given as a complete photoperiod, but in those groups where the light pulse was experienced between 11·5 and 16 hours after the onset of the primary photoperiod, a strong gonadal development was induced, whereas light pulses experienced at other times failed to stimulate any significant development.

The actual position of the photoinducible phase varies from species to species, and also intraspecifically in those species which might have a large geographical distribution. In *Coturnix*, the peak appears to be about 14 hours after dawn, but in house finches it occurs 12 hours after the start of the light period, and in juncos some 16 to 18 hours later. Light pulses given before or after these peaks are less effective and the limits of tolerance well marked. Other vertebrate groups have been less extensively studied, but a similar photoinducible phase has been shown in some mammals and also in at least one fish, the stickleback. In ferrets for instance, breeding can be induced in two months if animals are kept under an 18-hour photoperiod, but 6 hours of light per day are just as effective provided that 2 hours of it are given as an interruption to the dark period 18 hours from the start of the primary photoperiod that is, 12 hours after dark. If the primary photoperiod is extended to 12 hours the supplementary light pulse need only be given 6 hours after dark, indicating that the photoinducible phase in this species is phase-locked to the start of the light period and is irrespective of the length of the photoperiod. This, however, is not true in every group, since there is evidence in some insect species that the whole of the photoperiod may be significant in the entrainment of the photoinducible phase and by altering the length of the primary photoperiod a phase-shift can be induced.

In insects, the suddenness with which development is arrested in nearly all species once the critical photoperiod has been reached certainly points to the great accuracy of the time measurement involved and renders them particularly suitable for experimentation. One of the difficulties inherent in studying the biological clock in birds, for instance, is that the rhythm can only be demonstrated by examining a gonadal response which may take weeks to develop, but in insects, diapause, or its termination, develops very rapidly so that a response is sometimes manifest a few days after the stimulus. By applying suitable light-break techniques and analysing the incidence of diapause which these provoke, the photoinducible phase is as clearly delineated in many insect species as it is by the analysis of gonad response in vertebrate forms. For example when cabbage white butterfly larvae are reared under a normally strongly diapause-inducing 6L:18D day, with an additional 2-hour light pulse given as a light-break in the scotophase, the light-sensitive peak is revealed to occur about 16 hours after the beginning of the primary photoperiod (Fig. 6–4), so that in insects reared under these conditions (group 4, of Fig. 6–4) the light pulse tends to avert diapause induction. When the light pulse is experienced at other times during the scotophase it is less effective in preventing a diapause. In this species only one peak of light sensitivity occurs in the scotophase, but in some species dark scanning reveals two peaks when maximum diapause inhibition can be induced. The pink bollworm is an example of such a type. This moth has a larval diapause, and in light-break experiments the larvae demonstrate such a bimodal response (Fig. 6–5). A light

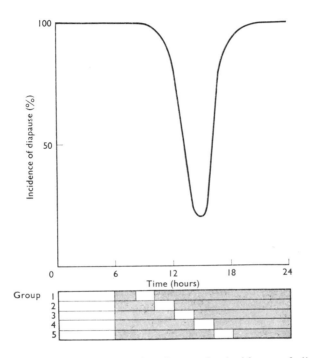

Fig. 6–4 Effect of 2-hour light-breaks on the incidence of diapause in cabbage worm maintained under a daily 6-hour photoperiod. (BÜNNING, E. (1964). *The Physiological Clock*, Academic Press, New York and London.)

interruption at either 15 hours or 20 hours after the onset of the primary photoperiod inhibits diapause, but when it coincides with the second phase it is always more effective in reducing the incidence of diapause. The time during the scotophase when maximum photosensitivity occurs is dependent on the length of the primary photoperiod. It seems that this insect may utilize both dawn and dusk as environmental 'cues' to synchronize the circadian oscillator and effect the measurement.

Circadian photosensitive oscillations do not seem to form the basis of all insect photoperiodic mechanisms. In the photoperiodically controlled aphid *Megoura viciae* the duration of the dark period is the critical factor which determines whether parent aphids become ovipara producers or virginopara producers. In their natural environment the short days of winter induce the production of the sexual, egg-laying females (oviparae), but the seasonal advent of long days cause their replacement by the parthenogenetic females. It is not, however, the duration of the light period

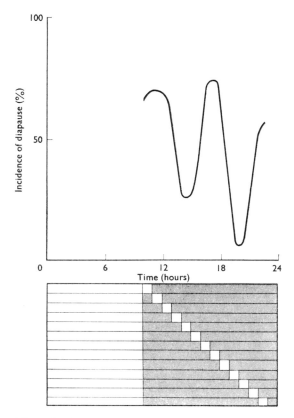

Fig. 6–5 Effect of 1-hour light-breaks on the incidence of diapause in pink bollworm maintained under a daily 10-hour photoperiod. (ADKISSON, P. L. (1964). *Am. Nat.*, **98**, 357.)

which determines this change, but the duration of the dark phase. Provided the dark period exceeds a duration of 9·5 hours ovipara production will nearly always take place, even though the light phase may be experimentally extended to over 30 hours. Thus the relevant part of a winter day is the long night. A short photoperiod coupled with a short dark period (8L : 4D) will not promote the development of oviparae, and it is only when the latter exceeds the critical length that this can be brought about. This suggests that the photoperiodic clock in *Megoura* measures the duration of the dark period and does not employ a light-sensitive oscillation. The scotophase has also been shown to be more critical than the duration of the photophase in the induction of diapause in several other species of insects.

6.3   Time-measurement and the seasonal cycle

The nature of the curves obtained from night interruption experiments outlining the photoinducible phase of the circadian light-sensitive rhythm, explains many of the properties of the response mechanism under natural

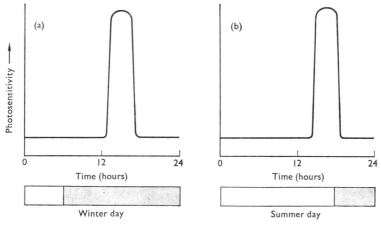

Fig· 6–6   Diagram to demonstrate how the photoinducible phase becomes engaged with the lengthening photoperiods of late spring days.

daily photoperiods, and indicates how such a rhythm can form the basis of the time-measuring mechanism which synchronizes the seasonal reproductive cycles or the seasonal induction of diapause. Thus in winter, the daily photoperiod is, for many species of birds, below the critical threshold and the photoinducible phase will always occur during the hours of darkness, as indicated in Fig. 6–6a. Therefore, stimulation of gonadotrophin hormone release does not occur and the animal remains sexually regressed. As daylengths increase in spring, however, the degree of coincidence between light and the photoinducible phase will increase causing a greater secretion of gonadotrophins and a consequent recrudescence of gametogenic activity (Fig. 6–6b). Eventually, complete coincidence will occur and the gonads will grow at a maximal rate. The critical photoperiod which produces a stimulatory response, will therefore, be determined by the precise position of the photoinducible phase in the 24-hour daily cycle of any particular species. Furthermore, this dictates the pattern of the annual reproductive cycle. In species where the photosensitive peak occurs early on in the daily cycle, it will be engaged earlier in the year and be manifest as an early breeding season. Thus, house finches with a circadian oscillation entrained to reach a photoinducible peak some 12 hours after dawn, will start gonadal development much earlier in the season than a species such as the junco where the photoinducible phase develops some 16 hours after dawn. In the latter case, the birds remain sexually quiescent until later in the year, the natural photoperiods having to become much longer to coincide with the later phasing of the photoinducible peak.

Photoperiodic Regulation of   7
Non-seasonal Cycles

7.1 Behavioural photoperiodism

In our consideration of the phenomenon of animal photoperiodism, discussion has been largely restricted to its role in the regulation of seasonal cycles. There is ample evidence, however, that, in addition to this, light also plays a regulatory role in the daily behaviour patterns that are apparent in the activities of most animals. All species whether they be sub-arctic, temperate or equatorial, show cyclic patterns of behaviour. These include such activities as locomotion, feeding, mating and oviposition, and the daily periodicity of these events are generally characteristic of any particular species. They can usually be classified into diurnal (daylight activity), nocturnal (night activity) or crepuscular (active during dawn or evening twilight) types.

The daily pattern of locomotory activity, and the feeding activities which are closely associated with it, have been extensively studied in a large range of both vertebrate and invertebrate land animals, and it is now well established that these rhythms whether they be diurnal, nocturnal or crepuscular, are primarily of an endogenous nature. When, therefore, such animals are exposed to conditions of continuous darkness or illumination, their activity rhythms become free-running and show a circadian periodicity. Under natural conditions, of course, circadian rhythms are never free-running but are synchronized to exactly 24 hours by their natural *Zeitgebers*, and although behavioural patterns are influenced by a variety of factors such as temperature, humidity or even in some cases by the sound of members of the same species, the primary entraining agent is nearly always the daily photoperiod.

In an animal kept under experimental conditions allowing the rhythm to free-run, reintroduction of a normal daily photoperiod will re-entrain the behavior to a precise 24-hour periodicity. This is very well exemplified in the motor activity of caged mice. Daily rhythms of locomotor activity are very easily studied by means of sprung or balanced cages in which the movement of the animal is recorded, either by the closing of an electrical circuit producing an impulse, or through a system of levers. In small caged rodents a very convenient technique is to record the running activity in an exercise wheel coupled to an electric counting circuit. The revolutions of the wheel can be recorded by a pen which is reset to the base line every minute. The height of the line is proportional to the number of revolutions of the wheel in the minute. Figure 7–1 shows such a record of the running activity of a white-footed mouse, *Peromyscus leucopus noveboracenis*, under different photoperiodic conditions, and illustrates the points discussed above. Each horizontal line represents a single 24-hour span, and 25 days' records are shown from the top to the bottom of the figure. During the first 7 days the mice were kept in constant darkness, and the record of their running activity clearly shows the circadian nature of this

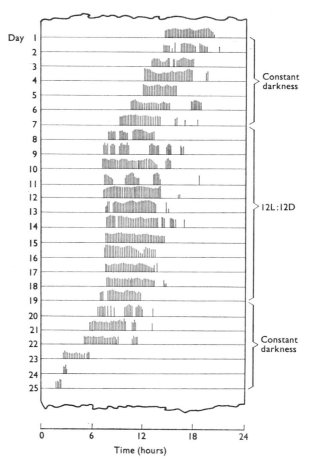

Fig. 7-1 Record of the running activity of a white-footed mouse maintained in constant darkness, and during a period of exposure to daily cycles of 12L : 12D.

rhythm. Thus, the daily onset of the motor activity became earlier each day. With the establishment of a 12-hour daily photoperiod on the 8th day, however, entrainment of the rhythm took place, and the onset of the daily activity become synchronized to start at approximately the same time. Replacement of the mice in continuous darkness on the 19th day caused the rhythm once again to 'free-run'. Many such rhythms have been recorded in birds, mammals and insects, and, although in some cases they may be simple with but a single peak of activity per day, bimodal responses are also quite common. In birds, for example, most of the patterns of behaviour such as nest-building, song or locomotor activity, both in diurnal and nocturnal species, show such a bimodal pattern, with a major peak at the beginning of the activity and a minor one towards its end. These can be related to dawn and dusk, and the same is true also of many insect rhythms.

Because an activity pattern is maintained in a constant time relationship with the photoperiod, by experimental manipulation of the photoperiod it is possible to re-entrain the rhythm so that it occurs at the reverse time to the periodicity displayed under natural conditions. Thus, a nocturnally active insect, for example, exposed to an artificially reversed photophase and scotophase will eventually readjust its daily activity pattern so that the peak now occurs during the artificial scotophase and is no longer coincident with the rhythm displayed under natural conditions. The number of cycles through which it passes before the rhythm is completely reset varies both with the species and also with the type of behavioural rhythm which is being investigated.

7.2   Photoperiodic regulation of insect emergence rhythms

In insects the emergence of the adult stage from its pupal or final nymphal form occurs during precisely defined times of the day, the actual time being dependent on the species. When a population is kept under observation, the succession of the daily production of adults is found to form an endogenous pattern which, like other such rhythms already discussed, shows a phase-shift when kept under continuous illumination or darkness. Similarly too, it is entrained by the photoperiod. *Drosophila* species, in particular, have been extensively studied in connection with this phenomenon.

In *D. pseudoobscura* adult emergence tends to occur principally at dawn or soon after, and the periodicity of the rhythm is entrained by the light experienced during the larval and pupal phases. The ultimate emergence of an adult is, of course, a reflection of the rate of developmental process during the preceding stages, and the entrainment of the photoperiodic emergence rhythm is, in effect therefore, a synchronization of developmental rhythms in the individual insects of a particular population. When *Drosophila* cultures are reared from the egg stage in continuous darkness, no well defined rhythm of adult emergence develops, but a single brief exposure to a light pulse during larval or pupal development is sufficient to act as a phase-setter and establish the rhythm. An electronic flash of only 1/2000 of a second is sufficient, and indicates that the onset of light, rather than the actual length of the photoperiod is the significant reference point for synchronization. Conversely, a light-off stimulus at the end of the photoperiod can also act as a *Zeitgeber*, and cultures reared under continuous illumination can have their emergence rhythm synchronized by a dark pulse, although under these circumstances the subsequent phase-setting differs from that entrained by a light pulse. It seems likely that in their natural environment since both types of cues (i.e. lights-on, lights-off) will be experienced daily, the flies make use of the temporal relationships between dawn and dusk, which in effect will give a measurement of the length of the photoperiod, in their synchronization of the rhythm. This has an adaptive significance in that it phases the emergence of adults to the cooler and more humid part of the day, and thus reduces the dangers of dessication to which the newly emerged image is particularly susceptible.

7.3   Photoperiodic effects on the oestrous cycle and ovulation

In non-seasonal polyoestrous species such as the laboratory rat, light can have an effect on the duration of the oestrous cycle. Under a normal 12 hour photoperiod the cycles in laboratory bred rats are usually of 4 to 5 days duration, but under reduced photoperiods or under conditions of continuous darkness, the dioestrous phase (period between the end of one oestrous cycle and the start of the next) of the cycle is lengthened and might eventually result in the production of a continuous anoestrous condition (period of sexual quiescence). Continuous illumination, on the other hand, leads to a state of persistant oestrus in which the ovaries maintain sets of graafian follicles, and the walls of the vagina remain cornified. Groups of rats exposed to different photoperiods, tend to go into a prolonged dioestrous phase if the duration of the light cycle falls below 10 hours whereas a photoperiod of over 14 hours increases the incidence of specimens showing a persistent oestrous condition.

Quite separate from the effect of light on the mechanism phasing the oestrous or seasonal cycles is its effect on the processes controlling the actual time of ovulation. In laboratory rats, repeated autopsies and histological analysis of the ovaries demonstrate that ovulation always takes place at night-time, usually between 1 a.m. and 3 a.m. This is a photoperiodically regulated phenomenon and by experimentally advancing or retarding the onset of the photoperiod the ovulatory rhythm will eventually re-entrain in response so that ovulation will occur earlier in the night or later, respectively. The ovulatory sequence in domestic hens can also be influenced in a similar manner, but whereas ovulation in the laboratory rat takes place in the dark, ovulation in the hen occurs in the daytime.

Further Reading

ASCHOFF, J. (1965). *Circadian Clocks*. North-Holland Publishing Company, Amsterdam.

BECK, S. D. (1968). *Insect Photoperiodism*. Academic Press, New York and London.

BÜNNING, E. (1964). *The Physiological Clock*. 2nd ed., Academic Press, New York and London.

CARTHY, J. D. (1966). *The Study of Behaviour*. Edward Arnold, London.

HARKER, J. E. (1964). *The Physiology of Diurnal Rhythms*. Cambridge University Press, London and New York.

LOFTS, B. and MURTON, R. K. (1968). Photoperiodic and physiological adaptations regulating avian breeding cycles and their ecological significance. *J. Zool. Lond.*, **155**, 327–394.

SOLLBERGER, A. (1965). *Biological Rhythm Research*. Elsevier, Amsterdam, London and New York.

WITHROW, R. B. (1959). *Photoperiodism*. American Association for the Advancement of Science, Publication No. 55, Washington, D.C.

WOLFSON, A. (1964). Animal photoperiodism. *Photophysiology*, **2**, 1–49.

References

BAGGERMAN, B. (1960). Factors in the diadromous migration of fish. *Symp. zool. Soc. Lond.*, **1**, 33–60.

BAKER, J. R. (1938). The evolution of breeding seasons. In *Evolution*, de Beer, G. R. (ed.), pp. 161–177. Oxford University Press, London and New York.

DANILYEVSKY, A. S. (1957). Photoperiodism as a factor in the formation of geographical races in insects. *Ént. Obozr.*, **36**, 5–27.

HARRINGTON, R. W. (1950). Preseasonal breeding by the Bridled shiner, *Notropis bifrenatus*, induced under light-temperature control. *Copeia*, **4**, 304–311.

HARRINGTON, R. W. (1957). Sexual photoperiodicity of the cyprinid fish, *Notropis bifrenatus*, in relation to the phases of its annual reproductive cycle. *J. exp. Zool.*, **135**, 1–47.

JENNER, E. (1824). Some observations on the migrations of birds. *Phil. Trans. R. Soc.*, **1824**, 11–41.

LEES, A. D. (1959). The role of photoperiod and temperature in the determination of parthenogenetic and sexual forms in the aphid *Megoura viciae* Buckton, I. The influence of these factors on apterous virginoparae and their progeny. *J. Insect Physiol.*, **3**, 92–117.

MAIRAN DE (1729). Observation botanique. *Hist. Acad. Roy. Sci. Paris*, **1729**, 35.

MARSHALL, A. J. (1955). Reproduction in birds: the male. *Mem. Soc. Endocr.*, **4**, 75–93.

MASAKI, S. (1961). Geographic variation of diapause in insects. *Hirosaki Daigaku Nogakuku Gakujutsu Hokoku (Bull. Fac. Agr. Hirosaki Univ.)*, **7**, 66–98.

ROWAN, W. (1925). Relation of light to bird migration and developmental changes. *Nature, Lond.*, **115**, 494–495.

ROWAN, W. (1926). On photoperiodism, reproductive periodicity and the annual migrations of birds and certain fishes. *Proc. Boston Soc. nat. Hist.*, **38**, 147–189.

ROWAN, W. (1929). Experiments in bird migration. I. Manipulation of the reproductive cycle: seasonal histological changes in the gonads. *Proc. Boston Soc. nat. Hist.*, **39**, 151–208.

ROWAN, W. (1930). Experiments in bird migration. II. Reversed migration. *Proc. natn. Acad. Sci. U.S.A.*, **16**, 520–525.

ROWAN, W. (1938). Light and seasonal reproduction in animals. *Biol. Rev.*, **13**, 374–402.

SCHÄFER, E. A. (1907). On the incidence of daylight as a determining factor in bird migration. *Nature, Lond.*, **77**, 159–163.